HIGH POINT UNIVERSITY

and the

FURNITURE INDUSTRY

RICHARD R. BENNINGTON

THE
History
PRESS

Published by The History Press
Charleston, SC
www.historypress.com

Copyright © 2021 by Richard R. Bennington
All rights reserved

Front cover, top: (left) Courtesy of High Point University; (center and right) Courtesy of International Market Centers; bottom: Courtesy of the High Point Market Authority.
Back cover, top: Courtesy of Richard Bennington; bottom: Courtesy of High Point University.

First published 2021

Manufactured in the United States

ISBN 9781467149242

Library of Congress Control Number: 2021937166

CONTENTS

CONTENTS

CONTENTS

ACKNOWLEDGEMENTS

I have been able to acquire quite a bit of knowledge about the furniture industry over the years, and I am grateful to a number of people who taught me about the industry and to others who provided ideas and materials that helped make this book a reality. First, Joe Meadors, retired from Bassett Furniture and my first teacher in the industry, and others who are working at or retired from Bassett who provided photographs and ideas that helped me tell a more complete story. Joe Carroll, retired publisher of *Furniture Today*, a good friend and a constant source of ideas and inspiration. Rhonda Lewis and a number of other current *Furniture Today* employees who were very helpful in pointing me in the right direction for photographs and other information. Tammy Nagem and her coworkers at the High Point Market Authority and Cindy Hodnett and her coworkers at the International Market Centers were all very helpful in securing photographs and other materials on the High Point Market. Former students, especially those who are named in the manuscript and are now working in the industry, provided photographs and other information.

There are also many people at High Point University (HPU) who helped me decide to write the book and provided assistance throughout the entire process. Dr. Jim Wehrley, dean of the Earl N. Phillips School of Business, gave me the original idea of compiling a timeline of what I have done since coming to High Point University and then of writing this book. Dr. Dennis Carroll, retired provost, encouraged me to start writing. David Bryden, HPU library director, helped put the pieces together and, most importantly, introduced me to Leanne Jernigan, a librarian in the Wanek Center who edited the manuscript and was the technical expert who helped me get the material organized properly, and Chelene Marion, the graphic designer who used her talents to design just the right cover for the book.

I am grateful to Dr. John Turpin, dean of the School of Art and Design, and Dr. Jane Nichols, chair of the Department of Design and Home Furnishings, for providing me a place to work in Norton Hall. And especially Benita VanWinkle, who was invaluable in using her ability to make the photographs appear just right in the book. Also, Martha Ashley, who answered my questions and helped me bring the manuscript to reality, and Cathy Nowicki, who encouraged me and provided information when I needed it. Chris Dudley and Gale Varner of the Institutional Development office and Barry Kitley, HPU vice-president for Community Relations, read the manuscript and helped me secure much of the information. Campus photographer Lee Adams helped me to locate High Point University photographs to make the book more complete and understandable.

There were others who read the manuscript and were helpful throughout the entire process. I am very grateful to them all for the help I received to be able to tell the story of the extraordinary partnership between High Point University and the furniture industry.

INTRODUCTION

When you see the title, you may think to yourself, "You have to be kidding. Why should I want to read a book about High Point University and the furniture industry together in one volume?" Perplexed though you may be, I suggest it is worthwhile reading; it presents a unique situation in which two distinctly different entities have come together in many ways to partner for the betterment of both. The story of HPU and its evolution alongside High Point's furniture industry puts a different spin on the relationship between "town and gown."

"Town and gown" is a phrase I have heard periodically during my professional life in higher education. I am not sure where this phrase originated, but most people take the word *town* to mean the community, town or city where an institution of higher education is located. *Gown*, on the other hand, is the college or university that just so happens to be located in the town. The academic community—the professors, students, counselors, resident advisors in the dorms, deans, the president, anyone working or studying there—are included under the overall term *gown*.

The common understanding seems to be that in places where there are colleges or universities, "town" and "gown" are considered two separate entities. In town, people are busy living their lives, working in factories, offices or whatever other types of businesses happen to be located in the area, sending their children to local schools, eating at local restaurants and attending church services. Often, this is done without much interaction with the college or university "gown," except to maybe attend athletic or other events where the town is "invited" to the campus. College or university students, for the most part, are just as insulated from the neighboring community; they live and eat on campus and socialize with their fellow students, perhaps their fraternity brothers or sorority sisters. In addition, what the students major in often does not have any direct connection with the employment opportunities in the local area. "Town and gown" are typically two distinct communities with relatively little interaction between them.

I have learned, however, that High Point is a rare exception to this rule; in fact, a vibrant partnership has developed between the university in High Point and the town's furniture industry—historically, the dominant business segment in the area. Little did I know when I moved to High Point over forty years ago to teach at High Point College that the two dominant influences on my life, other than my family and my church, would be the college (now High Point University) and the furniture industry. High Point College would be my employer for the next forty-plus years; in 1979, I took a leave of absence to work in the furniture industry, and on my return to campus, I became the first director of the new furniture marketing program.

It all began on August 9, 1974. I was driving a U-Haul truck from Athens, Georgia, to High Point, North Carolina, with my wife following behind in our family car. I was listening to the truck radio so I could hear the news. The rumor was that Richard Nixon might resign and that Gerald Ford would be sworn in as the next president of the United States. It was a seemingly quick trip because there was nonstop news from Washington, D.C. To make a long story short, when we left Athens, Richard Nixon was president, and when we arrived in High Point, Gerald Ford was president. I had just completed my graduate program and received my doctorate from the University of Georgia and had accepted a position to teach economics and business at High Point College. When we moved into our new home just off the High Point College campus, I had no idea where this move to a new college in a new city would take me.

Actually, my first introduction to the furniture industry began when I was just a young boy. I was raised on a general livestock farm in the mountains of southwestern Virginia. I enjoyed farm life, but some of my favorite memories were our monthly trips to Mountain City, Tennessee, to visit my mother's dad and stepmother. During these visits, I enjoyed watching people come to make purchases in my grandfather's retail store, where he sold household furniture, pianos and appliances—and, oh yes, televisions. They were an especially hot item in the 1950s. My grandfather was a talented pianist, and by training, he was a piano tuner and repairman. He always kept a piano

Richard Bennington and his grandfather R.J. Eastridge outside of his furniture store in Mountain City, Tennessee. *Author's collection.*

in the front window of the store. In the summers, when there were very few customers in the store, he would open the front door and begin playing the piano. Little by little, people would begin to drift into the store to listen to the music, browse through the products and, hopefully, see something they might want to buy. He also made free promotional items, like calendars with the store's name and phone number on them, available to the visitors. I guess this was my first introduction to furniture marketing. Perhaps it was natural that when I was taking a marketing course in graduate school, I chose "Marketing in the Retail Furniture Industry" as the topic for one of my term projects.

Before moving to High Point, North Carolina, the main thing I knew about the city was that my uncle who owned a wood furniture components factory regularly came to the High Point Furniture Market to visit his customers and see how his components looked on the products in the market showrooms. As I drove around High Point and the neighboring town of Thomasville, I began to see brand names on factories and showroom buildings that I recognized from my grandfather's store. The sheer size and number of furniture factories and market showrooms was impressive, and I soon realized I had truly moved to the "furniture capital of the world."

My first teaching responsibilities were economics and business courses. Since I was the "new guy on the block," I was assigned to teach a wide variety of business courses, such as Principles of Economics, Principles of Accounting, Principles of Management, Principles of Marketing and even Personal Finance. I enjoyed the interaction with students and the challenge of teaching so many different types of classes. As I began my teaching career at High Point College, I started to learn about the impact that the furniture industry had on the city of High Point. It also dawned on me that there were connections between High Point College and the furniture industry. I checked out books from Wrenn Library and attended services at Hayworth Chapel, both buildings that carried "furniture family" names. Furniture executives were among the members of the college's board of trustees. Business graduates were employed by furniture companies in areas such as management, sales and finance. Chemistry graduates were employed by furniture manufacturers in their furniture finishing departments and companies that manufactured the furniture finishing materials.

The more I drove past furniture market showrooms like the Southern Furniture Exposition Building on Main Street, factories where various types of furniture were being manufactured in the southern part of High Point and nearby Thomasville, I found myself wanting to know more about the furniture industry. I met people involved in the furniture industry, such as Darryl and Stella Harris, who, at that time, were operating a retail store named Furnitureland South located on High Point's South Main Street. I talked with others like Norman Silver and Bob Dutnell, who were involved in the management of Silvercraft, an upholstered furniture manufacturing company located in southeastern High Point. The more I was told about their companies and the products they made or sold, the more I wanted to know.

My desire to learn more about the furniture industry became a reality in 1978, when an initiative was launched to determine how High Point College could help meet the marketing education needs of the home furnishings industry. The initiative resulted in a series of meetings that were held to obtain feedback from many of the key leaders of the home furnishings industry. Included were representatives of manufacturers, retailers, industry suppliers and industry trade associations. The conclusion was that there was a definite need for a college-level educational program in furniture marketing, and because of its location in High Point, North Carolina, with its large number of furniture manufacturers and other furniture-related businesses, as well as the world's largest wholesale furniture market, High Point College was deemed the ideal place to offer this and other such educational programs. Because of my interest in learning about the industry, I agreed to take a six-month leave of absence from High Point College and work full time in the industry, so that I could help formulate a major in Furniture Marketing that would meet the needs of the industry.

During that six-month period, I spent most of my time working with Bassett Furniture Industries, which, in the 1970s, was the largest household furniture manufacturer in the world. I worked full time for several weeks in the Bassett Furniture factories, rode with one of their sales representatives as he made sales calls to retailers throughout his sales territory and experienced the entire furniture marketing cycle from product development through seeing how the products were presented to the retail buyers in their market showrooms. Since that time, I have had the opportunity to visit retail stores, spend time in other factories and visit furniture market showrooms in High Point and other markets around the world. This has given me insight into the dynamic, ever-changing furniture industry.

I also learned that there is both a business side and a human side of the industry. Managers of furniture companies are fierce competitors, but at the end of the day, they meet and treat each other like friends. They love to socialize and go to industry events together. They also develop friendships with their customers, especially between manufacturers and retailers. In essence, the historical roots of the industry seem to be related to fair dealing, a concern for both customers and competitors and a genuine love for the products themselves. Above all else, they love the industry.

I think the human side of the business is also revealed through industry events like two that are held during the High Point furniture markets: the Furniture Fellowship Prayer Breakfast, which is held during each spring furniture market, and the annual American Furniture Hall of Fame induction banquet, which is held during each fall furniture market. These events, described in chapter 11, are only two of the many examples that serve as evidence that the same values High Point University works to instill in their students are practiced in the furniture industry. Perhaps it is because of the nature of the industry that these two entities share common values. High Point University is working hard to develop students who will lead lives of significance, to be hardworking, productive individuals and good citizens. And the industry is working hard to identify consumer needs and wants so that it can provide the home furnishings products to meet them.

High Point College opened in Furniture City USA in 1924, and when I moved there fifty years later, an unprecedented synergy between "town and gown" was already apparent. It began slowly as part of a civic endeavor to convince the Methodist Church to locate a college in High Point, but over the long term, the relationship has developed into a mutually beneficial partnership. I am eager to share what I have learned in my forty-year journey of involvement with High Point University and the home furnishings industry. I also think you will enjoy what I have learned through interviewing High Point College/University graduates who now work in the furniture industry. I have discovered that they are doing some amazing and interesting things.

Chapter 1

HOW IT ALL BEGAN

High Point Furniture Manufacturing, High Point Furniture Market and High Point College

In order to understand the partnership between the furniture industry and High Point University, we need first to explore the historical roots of the city of High Point and uncover how it came to be known as "Furniture City." This chapter will trace the early onset and evolution of furniture manufacturing in and around High Point through the emergence of what was originally called the Southern Furniture Market. It was in the context of this early thriving economy that the city was able to bid for—and win—the prospect of High Point College opening on city soil.

FURNITURE MANUFACTURING IN HIGH POINT

Several years ago, my daughter bought me an attractive framed print of an antique map of the city of High Point for my office at High Point University. The title on the map reads: "Aero View of High Point, North Carolina, The Manufacturing Centre of the South, population 12,000. Published and copyrighted in 1913 by J.J. Farris of High Point." The map is intricately hand drawn, shown as if someone was looking down on the city of High Point from the north. The main streets are carefully drawn, as are the downtown area, public schools, a number of the residential areas and the railroad running through the center of town. The entire south side of the map contains images of a large number of factories, each with its tall chimney billowing forth a wispy column of smoke. The major buildings on the map,

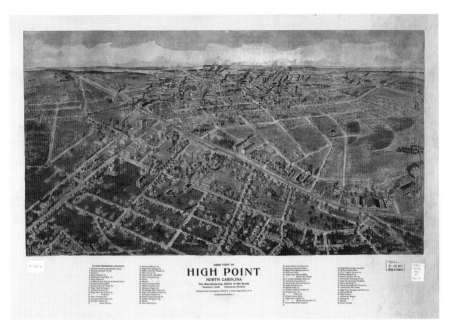

Aerial view of High Point, North Carolina, the manufacturing center of the South, circa 1913. *Courtesy of J.J. Farris of High Point.*

including the factories, are numbered, and their names are carefully listed at the bottom. The list contains a wide variety of factories, such as cotton mills, a hosiery mill, a buggy factory, a couple of casket manufacturing companies and, as would be expected, a bunch of furniture factories.

A column, written in October 2013 for *Furniture Today*, the industry's leading trade publication, by its former publisher Joe Carroll, titled "Mighty Oaks from Little Acorns Grow" sheds light as to how furniture manufacturing moved to High Point. Carroll wrote that in 1867, William Henry Snow moved his family from Vermont to Greensboro, North Carolina, in search of a warmer climate and the availability of less expensive timber for his manufacturing business, which was making the wood shuttle blocks and bobbins used in weaving by textile mills. It wasn't long before his company became the world's largest manufacturer of shuttle blocks, and local businessmen were asking Snow to set up his factory in High Point. Shortly thereafter, the move to High Point was complete, and Snow expanded his business to include making wagon wheel spokes and axe and hammer handles.

In addition to making and selling these products, he apparently also developed a lucrative business selling lumber. While on business trips to sell his lumber, Snow noticed that there seemed to be a good opportunity for a company to make and sell inexpensive wood furniture. While returning from one of these trips, he met with John H. Tate and Thomas F. Wrenn, who were prosperous grocers at the time. They each invested $1,000, and the High Point Furniture Company was born. The factory opened in 1869 with twenty-five workers. Snow furnished the lumber, and High Point Furniture Company made the finished products. By the end of the company's second year, it employed eighty-five people and made twenty different styles of bedroom furniture.

Just as the furniture industry had transitioned from New England and Upstate New York to Grand Rapids and other locations in the Midwest, manufacturers began to see the advantages of relocating to High Point, with its cheap labor, abundance of nearby raw materials and easy access to rail transportation, which provided companies the ability to ship products relatively easily to much of the eastern United States. Therefore, by 1913, when the map was drawn, High Point had truly become a center of furniture manufacturing.

To fast-forward to 1923, the year before High Point College opened, the city directory lists twenty-five furniture factories producing a wide variety of furniture products.

Examples were:

- Alma Furniture Co., named for the daughter of one of the company's cofounders, produced wood office desks.
- Giant Furniture Co. produced chamber suites (consisting of dressers, beds and chiffoniers).
- Globe Parlor Furniture Co. was recognized for its "highest class" couches, lounges, davenports, library suites, rockers and parlor suites. (It also made "handsomely designed" world globes.)
- High Point Furniture Co. was known as a volume producer of moderately priced bedroom furniture. It is said that High Point Furniture Co. was the first furniture manufacturer to ship a "solid carload" (railroad car) of furniture from High Point.
- Ideal Table Co. produced library tables, extension tables and dining suites.
- Kearnes Upholstering Co. produced medium-grade popular suites, lounges and couches.
- Marsh Furniture Co. produced knock-down kitchen safes (to be assembled by the customer), along with center tables and hall racks.

Tomlinson Chair Manufacturing Co. *Courtesy of the High Point Historical Society, High Point Fire Department Collection.*

- Tomlinson Chair Manufacturing Co. pioneered the production of totally coordinated upholstered furniture groupings. (Tomlinson was also credited with being the first to create a showroom where retail store buyers could select products to sell in their stores.)

From reading the early product descriptions, it is obvious that terminology has changed since the 1920s. To cite a few examples: the term *chiffonier* has now been replaced with the term *chest of drawers* (especially in referring to bedroom suites); center tables are now called coffee tables; couches and davenports are just called sofas; and the term *chaise lounge* has replaced the more generic term *lounge*. Even with the use of different terminology, this short listing of some of the twenty-five furniture manufacturers operating in High Point in 1923 makes it evident that most categories and quality levels of furniture were being produced in the city.

THE HIGH POINT FURNITURE MARKET

High Point was also establishing itself as a home for a wholesale market for furniture. The first market, largely showing products made by manufacturers of inexpensive wood furniture, was held in 1909. The popularity of the High Point Market gradually increased so that by 1913, the market was established as a semiannual event, with furniture being shown to retail buyers in storefronts, factories and warehouses. The first relatively large furniture showroom building in High Point, the Southern Furniture Exposition Building, opened in 1921 with 149 exhibitors. By 1924, the Southern Furniture Exposition Building was completely leased and other manufacturers placed on a waiting list.

Southern Furniture
Exposition Building.
*Courtesy of International
Market Centers.*

THE CITY OF HIGH POINT WINS THE BIDDING WAR
FOR A NEW METHODIST COLLEGE

Dr. Richard McCaslin, in his 1995 history about High Point University, *Remembered Be Thy Blessings, The College Years: 1924–1991*, recounts how the Methodist Protestant Church struggled for nearly a century to establish a college in North Carolina. Most of its efforts had targeted the more remote areas in the state in an attempt to keep the students away from the "evils" of city life. In 1861, the first Methodist Protestant college, Yadkin College, was chartered and opened in rural Davie County, North Carolina. Although the school's efforts may have been noble, its success seems to have been hampered by the sparse population in the area. And it soon became obvious that Yadkin College's prospects for survival were not very bright.

By combining Dr. McCaslin's research with information assembled by the High Point University library staff for the 2018 University Founders Day celebration, it is apparent that the person who was perhaps most passionate

Right: Reverend Joseph McCulloch. *Courtesy of High Point University Archives.*

Below: Roberts Hall, which was opened in 1924. *Courtesy of High Point University Archives.*

Opposite, top: Banner from *Methodist Protestant Herald*. *Courtesy of High Point University Archives.*

Opposite, bottom: The Yadkin College bell, which was moved to High Point College when Yadkin College closed. *Courtesy of Richard Bennington.*

about the founding of what was ultimately High Point College was Joseph F. McCulloch. An ordained Methodist minister and Guilford County native, McCulloch returned to Greensboro and purchased the equipment of a local newspaper, the *Central Herald*. Using this equipment, he started publishing what was called the *Methodist Protestant Herald*, and it was through this publication that McCulloch passionately advocated for the founding of a new Methodist Protestant college in the area.

Apparently, the efforts of McCulloch and others were successful because by the early 1920s, the decision was made to focus on establishing a new Methodist Protestant college in a more heavily populated area of the state. Three bids were submitted: one from Greensboro, a second from High Point and a third combined bid from Burlington and Graham.

High Point's offer of $100,000 and sixty acres of land was accepted in 1921. Following the construction of the first three buildings, High Point College opened in 1924. It is ironic that as the first students entered High Point College, Yadkin College closed. The Yadkin College bell was brought to the High Point College campus and is displayed near what is now Finch Dormitory today.

Although it was not obvious at the time, the opening of High Point College would provide fertile soil into which seeds would be planted, leading to a vibrant, mutually beneficial partnership between the furniture industry and the new Methodist Protestant institution of higher education in the city of High Point.

Chapter 2

THE GREAT DEPRESSION, WORLD WAR II, POSTWAR BOOM AND THE TOWN GOWN PARTNERSHIP SLOWLY BEGINS TO GROW

T he period between 1924 and most of the 1970s reflected the development of a relatively normal relationship between High Point College and the furniture industry, the dominant industry segment in High Point. By this, I mean that it was a rather typical relationship between town and gown. The trustees were involved in trying to raise the money to fund the college on a long-term basis, and the executives of the furniture companies were concentrating on their jobs in making and marketing furniture. A number of the High Point trustees were also either active or retired furniture executives, and their motivation for serving seems to have been civic in nature. It was simply a good idea to have a prosperous liberal arts college in the city of High Point. Citizens of the area could send their sons and daughters to High Point College for a good education, and companies in the area would have a local source of good employees. And it was during this time that the first seeds of cooperation or partnership were planted between the furniture industry and High Point College.

THE GREAT DEPRESSION AND WORLD WAR II AFFECTED BOTH THE FURNITURE INDUSTRY AND HIGH POINT COLLEGE

The household furniture industry is what economists call a "cyclical industry." This means that, overall, product sales vary with the business cycle. When business is in a prosperous part of the cycle, the furniture business is generally good. But furniture is a purchase that, for the most part, can be postponed. Therefore, when the business cycle is in a downturn or depression, the consumer can use what they have for a little longer, or, to use today's consumer as an example, they may buy something from a secondhand store or thrift shop to use until times are better, when they can buy something new that they would really like to have.

To get a basic understanding of the impact of the Great Depression and World War II on the furniture industry, let's look at the companies that were manufacturing furniture in the United States, especially in the High Point area, and at the High Point Furniture Market. Then we will look at High Point College and how it was affected by the Great Depression and World War II.

Impact on the Furniture Industry

The furniture industry, as well as other kinds of businesses in the United States, was adversely affected by the Great Depression and the United States' involvement in World War II. The high levels of unemployment during the early 1930s meant that consumers could not afford to buy home furnishings. Furniture manufacturing increased slightly during the mid- to late 1930s, but that increase didn't last long. Just as things began to improve long enough for the manufacturers to think positively about their businesses, World War II broke out in Europe, and the furniture business declined again.

During World War II, a number of furniture companies used their factories to support the war effort. For example, Thomasville Furniture Industries' plants were very active producers of bunk beds, plywood for glider planes and commercially sized kitchen supplies from rolling pins to huge spatulas for stirring large containers of food being prepared for the troops. Also, following the Japanese attack on Pearl Harbor, approximately 25 percent of the company's workforce answered the nation's call for service. This resulted in Thomasville Furniture, like many other companies around the country, hiring women for the first time in production jobs in order to keep its plants operating.

The furniture market was also affected by the Great Depression but to a greater extent by World War II. Beginning in 1942, thirteen and a half of the fourteen floors of the Southern Furniture Exposition Building in High

Products produced by Thomasville Furniture Industries. *Courtesy of Thomasville Furniture Industries.*

Wartime products included large wooden spatulas, tent stakes and rolling pins which were nearly 2 feet long.

Point, now part of the International Home Furnishings Center, were used to store personnel records for the military. The only furniture market during the war years was held in 1943, when the southern furniture manufacturers attempted to hold a partial market.

IMPACT ON HIGH POINT COLLEGE

Needless to say, the Great Depression adversely affected both the enrollment and financial stability of High Point College. To refer again to Dr. Richard McCaslin's 1995 history of High Point University *Remembered Be Thy Blessings*, Dr. McCaslin discusses in considerable detail the serious

World War II U.S. Army Air Corps training on the High Point College campus. *Courtesy of High Point University Archives.*

financial condition and low enrollment suffered by the college during the 1930s. Nonetheless, the institution was able to continue its operations and weather the Great Depression.

World War II also affected the normal operations of the college, with some of the dormitory facilities being used to house military cadets and the campus being used by an army air corps training detachment. This seems to have partially offset the decline in enrollment resulting from male students enlisting in the military and the departure of female students to accept jobs in business.

POST–WORLD WAR II HOUSING BOOM

After World War II ended, furniture manufacturing began to increase dramatically, fueled by a postwar housing boom. To use a quote from Dr. David Thomas's "A History of Southern Furniture," reprinted in the October 12, 1967 issue of *Furniture World* magazine, "After four years of deprivation, the American public was eager for an opportunity to spend money saved from high wartime income." This attitude resulted in a dramatic increase in the

Thayer Coggin Upholstered Furniture Factory. *Courtesy of Thayer Coggin Furniture.*

number of new homes being built in the United States. Many of these homes were being built with the help of mortgages guaranteed by the federal government. Servicemen returning from World War II were able to move into "GI houses" without making a down payment and with low monthly payments.

As a result, times were good for furniture manufacturing in and around High Point. A significant number of new furniture factories opened in the area, and by 1955, almost half of the wood bedroom furniture in the United States was produced within a 125-mile radius of High Point.

Companies that supplied the materials used in wood furniture manufacturing were also flourishing. Companies like Lilly, Guardsman and Mobil had large plants in the western section of the city of High Point, producing the lacquers, varnishes and other finishing materials needed in wood furniture manufacturing. Hayworth Roll and Panel of High Point was supplying plywood panels with a variety of fancy-face veneers for local manufacturers. Commercial Carving Company in Thomasville, one of the world's largest turning mills, provided turned legs and various carved wood components to furniture manufacturers in North Carolina and elsewhere.

Times were also good for upholstered furniture manufacturers, as well as manufacturers of other types of furniture. There were a number of successful company startups like Thayer Coggin, which produces quality upholstered furniture and is well known for its mid-century modern style products (see the image above). Suppliers such as Leggett and Platt, which supplied components such as polyurethane foam and metal springs for upholstered furniture, were experiencing prosperous times. Companies manufacturing the decorative upholstery fabrics used by the local manufacturers were also experiencing increased sales and built plants in and around High Point.

MARKET REOPENS AFTER WORLD WAR II

The first furniture market after World War II was held in January 1946 in Chicago at the American Furniture Mart, at the time, the largest furniture showroom building in the United States. Also in 1946, the army left the

Southern Exposition Building, and after renovations on the building were complete, the first "Southern Exposition" after World War II was held in High Point in January 1947. Fueled by the anticipated sales of furniture caused by the post–World War II housing boom, retail store buyers flocked to North Carolina to buy furniture for their stores.

The "Southern Market" Begins to Show Dominance

In 1948, the High Point Market set new attendance records, and construction began on expanding the number and size of showrooms in High Point. A year later, in 1949, televisions were shown at the market for the first time. And by 1950, even Kroehler Furniture Company of Naperville, Illinois (a suburb of Chicago), at that time the largest furniture manufacturer in the world, began showing in High Point.

Over the next twenty years, the numerous expansions to the Southern Furniture Exposition Building and other showroom buildings opening in High Point reflected the shifting of attendance from markets such as Chicago, New York and Dallas to High Point, making the "Southern Exposition" the leading home furnishings market in the United States.

Wrenn Donates the Wrenn Memorial Library

There are a few examples from the college's early days in which furniture executives or individuals from furniture families took an initiative to support the college in a meaningful way. One particularly strong example is Louise C. Wrenn, the widow of M.J. Wrenn, a prominent local furniture company executive, who donated the funds for the M.J. Wrenn Memorial Library, which opened on the High Point College campus in 1937. Wrenn was, at one time, the sole owner of High Point Furniture Company, the first furniture factory in the city of High Point, and half-owner of the Wrenn-Columbia Furniture Company, another early furniture manufacturer in High Point. M.J. Wrenn was also a former mayor of High Point and an early supporter of High Point College.

Wrenn Memorial Library. *Courtesy of High Point University Archives.*

Dr. Cummings Senses a Furniture Industry Need

Dr. Edmund O. Cummings, chairman of the Chemistry Department, was perhaps the first High Point College professor to attempt to help solve a furniture industry problem. As early as 1945, he began to offer a course in industrial chemistry to provide an overview of finishes for the wood furniture industry. Apparently, interest in this area of study increased until 1964, when High Point College opened a chemical coatings center within the Chemistry Department, with "Polymer and Coatings Chemistry," the key course in understanding furniture finishes, as a required course. The center's objective was to train chemists to work with wood furniture manufacturers in the local High Point area.

Another source of employment for the High Point College students were the companies that manufactured finishing materials for the furniture industry. This evolved into an early example of cooperation between High Point College and the furniture industry; senior students conducted research in both campus laboratories and on-site laboratories of area businesses.

Left: Dr. Edmund Cummings. *Courtesy of High Point University Archives.*

Below: Dr. Cummings Industrial Chemical Lab. *Courtesy of High Point University Archives.*

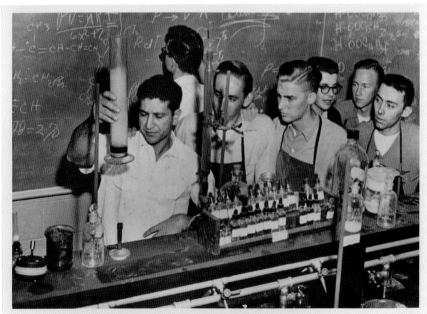

View of chemistry lab with students

To better understand the relationship between the High Point College Chemistry Department and the furniture industry, I interviewed Jim McGee, a High Point resident who graduated in 1951 with a degree in chemistry. McGee said that Dr. Cummings was particularly interested in paint and other coatings. In fact, McGee indicated that Dr. Cummings had obtained a number of patents for paint pigments and paints. The chemistry students were exposed to the real world in a variety of ways. For example, McGee and other students helped Dr. Cummings outfit a laboratory building he owned in the vicinity of High Point for conducting his private experiments. This was a forerunner to what was later called "experiential learning."

McGee's orientation to the furniture industry came from the fact that he and his fellow students got firsthand experience by spending a large part of their industrial coatings class working in the laboratories of the local plant of the Marietta Paint and Color Company, a furniture finishing manufacturer with a plant in High Point. From working in these labs, the students learned how paint was made and how wood furniture finishes were blended to meet the requirements of a certain furniture style or to accent the grain characteristics of a particular type of wood.

When he graduated, McGee went to work for the U.S. Atomic Energy Commission (AEC). After five years with the AEC, he returned to High Point and went to work with Lilly Industries Inc., another supplier of furniture finishes. His chemistry degree and the firsthand experience he had as a High Point College student were both invaluable in his new job. He worked for Lilly from 1956 to 1970, when he left to join Thomasville

Furniture Industries as assistant finishing superintendent. There, he worked under Thomasville finishing superintendent Wilfred Lamar, a 1940 graduate of High Point College. When Lamar retired, McGee became the Thomasville finishing superintendent and worked in that capacity until he retired.

Jim McGee, a High Point College chemistry graduate and retired Thomasville Furniture Industries finishing superintendent. *Courtesy of Richard Bennington.*

Dr. Cummings, his chemistry students and graduates like Jim McGee are good examples of how High Point College saw a need that was present in the local furniture industry and provided a solution. The partnership between town and gown was beginning to blossom.

FINCH LECTURE SERIES

The Finch lectures were made possible by a gift in 1959 to High Point University by the Charles F. Finch Foundation of Thomasville, North Carolina. The Charles F. Finch Foundation made the gift for the benefit of the Department of Religion and Philosophy at High Point College to set up the Charles Franklin Finch Lecture Series. The purpose of this lecture series was to foster dialogue and deepen the understanding regarding issues of importance to communities. Each year, this endowment makes it possible to bring an outstanding churchman, humanitarian or scholar to the High Point University campus.

The Finch name became associated with the furniture industry when Charles F. Finch and his brother T.J. Finch bought Thomasville Chair Company in 1907. By 1913, Thomasville Chair Company had grown to occupy twelve acres of land and produce 1,300 chairs per day. Ultimately, the company transitioned into a much larger company with a broader product assortment, Thomasville Furniture Industries.

THE CHARLES E. HAYWORTH SR. MEMORIAL CHAPEL

One of the prominent furniture families in High Point was the Hayworth family. The father, Charles E. Hayworth Sr., and his two sons, Charles E. Hayworth Jr. and David R. Hayworth, were involved in a number of furniture-related businesses in High Point. Some of the principal businesses

Charles Hayworth, Jr.

HPU Benefactors

David Hayworth

Charles Jr. and David Hayworth, High Point University benefactors. *Courtesy of High Point University.*

Charles E. Hayworth Sr. Memorial Chapel. *Courtesy of High Point University.*

they had were Hayworth Roll and Panel, the first plywood factory in the city; Myrtle Desk Company, a manufacturer of wood desks and library furniture; Alma Desk Company, at one time, the world's largest wood desk manufacturer; and Clarendon Industries, a manufacturer of office seating. The Hayworth brothers, Charles Jr. and David, followed Wrenn's example by becoming another furniture family who offered significant financial support through providing lead gifts for buildings on campus.

Charles E. Hayworth Jr. donated funds for a chapel on campus named in honor of his father, Charles E. Hayworth Sr. This seems to have been the first freestanding chapel on the campus and was named the Charles E. Hayworth Sr. Memorial Chapel. This chapel became the second major building in the first fifty years of High Point College's history to bear a furniture family name. The Charles E. Hayworth Memorial Chapel, which opened in 1972, remains the center of religious life on campus.

Chapter 3

THE FURNITURE INDUSTRY ASKS FOR INDUSTRY-SPECIFIC EDUCATION, AND I GO OUT TO SEE WHAT IS NEEDED

I n 1978, the partnership between the furniture industry and High Point College took a giant leap forward. That year, an initiative was begun to determine how High Point College could help meet the marketing education needs of the home furnishings industry. Heading the initiative was Charles E. Hayworth Jr., chairman of the High Point College Board of Trustees (he was also president of Alma Desk Company), and future board of trustees member Robert Gruenberg, who was general manager of the Southern Furniture Exposition Building, the largest furniture market showroom building in High Point. A third executive, Irvin Black, chairman of the Alderman Company, a large furniture photography and integrated marketing services company, also played a key role in advocating for the establishment of a furniture-related education program at High Point College.

LET'S MEET AND SEE HOW WE CAN PARTNER TOGETHER

Later that year, a series of meetings were held with a number of the leaders of the home furnishings industry. The individuals included in these meetings were key management officials from manufacturers such as Thomasville Furniture Industries, Bassett Furniture Industries, Broyhill Furniture

Bassett Furniture corporate headquarters in Bassett, Virginia. *Courtesy of Richard Bennington.*

Industries, Norwalk Furniture, Henkel Harris, Casard Manufacturing Company and Woodmark Originals and retailers like Haverty Furniture, Heilig Meyers and Kittle's Furniture. Also present were representatives from industry trade associations, including the Southern Furniture Manufacturer's Association (SFMA), National Home Furnishings Association (NHFA), International Home Furnishings Representatives Association (IHFRA) and National Wholesale Furniture Association (NWFA), and industry suppliers such as Erath Veneer Company.

The conclusion was that there was a definite need for a college-level educational program in home furnishings marketing but that it should be based on extensive research so it would be clearly focused on the needs of the industry. Because of its location in High Point, North Carolina, the geographic center of the home furnishings industry and home to the world's largest wholesale furniture market, High Point College was the logical place to offer educational programs in marketing for the home furnishings industry. This realization was a major milestone in forging a win-win partnership between High Point College and the furniture industry.

Several proposals were made as to what should be required in a furniture marketing program at High Point College. The result of the discussion was that an in-depth study should be undertaken to decide what specific courses and course materials would be most important for the training of students who, on graduation, would successfully fill meaningful positions in the furniture industry. Since I had been at High Point College for almost five years at that point and had been wanting to learn more about the furniture industry, I agreed to take an extended leave of absence from my duties as chair of the Earl N. Phillips School of Business so I could experience the industry firsthand. My goal was to gather the knowledge necessary to formulate a truly relevant major in furniture marketing.

An Outsider Is Shown the Ropes

Although I didn't want to admit it, I was an outsider who didn't have a good grasp of the products being sold and how the industry really worked. To remedy this, I worked with the management of Bassett Furniture Industries at their home office in Bassett, Virginia, for the first six months of 1979 in order to understand the unique aspects of furniture products, as well as gain a firsthand knowledge of the internal workings of the industry. Fortunately, one of the Bassett plant managers had a garage apartment in the town of Bassett where I could stay when I didn't want to drive the sixty miles back to my home in High Point. I was assigned to work with Joe Meadors, Bassett's senior vice-president

Joe Meadors, senior vice president of sales of Bassett Furniture Industries. *Courtesy of Bassett Furniture Industries.*

of sales. Meadors made sure that I experienced as much of the furniture industry as possible during my six-month leave of absence and remained a valuable industry contact afterward.

PRODUCT/MANUFACTURING SIDE OF THE BUSINESS

My first assignment was to spend a considerable amount of time experiencing the manufacturing of wood household furniture. To accomplish this, I rotated through the various departments of the W.M. Bassett Plant, a fully integrated "case goods" plant in nearby Martinsville, Virginia. Since I was new to the industry, I really wasn't sure what the term *case goods* meant. As I inquired as to the meaning of this term, I was given two origins. One was a reference to promotional wood furniture that was produced in the early 1900s. This furniture was shipped, usually by rail, in wood cases. Some of the people in the furniture industry, especially from the higher-priced manufacturers, jokingly said that the cases were so much better made than the furniture inside them that they just called the furniture "case goods." I soon found out that the real meaning referred to wood furniture that was sold in coordinated suites or sets (bedroom and dining room) that contained pieces with internal storage, such as china cabinets for dining room suites and dressers and chests of drawers for bedroom suites. These storage pieces were simply referred to as "case" pieces, with the complete suites being called "case goods."

Howard White, manager of the W.M. Bassett Plant, seemed to be a virtual walking encyclopedia of everything to do with wood furniture manufacturing. During my time at the plant, he made sure I learned as much as possible about the various processes the products went through during the manufacturing cycle. I learned that the plant's production was scheduled based on "cuttings." *Cutting* is the term used to refer to a quantity of products, usually of the same style, moving through the plant at the same time. For example, one cutting might be comprised of two hundred oak traditional-style bedroom suites.

I began in the "rough end," which is the area where raw lumber from the lumber yard was brought into the plant, cut and the machine processing began. The boards were sent through the planer so that both of the flat sides would be smooth. They were also sent through the rip saws so they would be the right width, and they were sent through the cutoff saws so they would be the right length.

W.M. Bassett Case Goods
Factory in Martinsville,
Virginia. *Courtesy of Bassett
Furniture Industries.*

Next, I learned that Bassett used veneers (thin layers of wood that varied in thickness from one-eighth of an inch to one-thirty-second of an inch), and I progressed to the veneer room, where wood veneer was received and cut to desired sizes. Then I saw panel (plywood) construction, where the fancy-face veneers were applied to the exterior sides, or "faces," of the panels, and cheaper veneers were applied to the back sides, or insides, of the panels for the wood bedroom and dining room products. These products were referred to as "all wood" products rather than "solid wood" products.

From there, I followed the products through the woodworking machine room, where the actual furniture parts were made, and then I went on to the sanding room, where the parts were sanded smooth. I watched the workers in the cabinet room, where the products were actually assembled. Next came the finishing room, where the decorative wood finish was applied to the products. In many of these areas, I was treated like a management trainee and actually participated in the production process. The last step was the inspection and packaging of the products for one of three destinations: storage in the Bassett Furniture warehouse, immediate shipment to retailers for retail floor sales samples or as finished goods inventory ready for sale to their customers.

Although Bassett Furniture had been known throughout most of its history as a wood furniture manufacturer, management had decided in the early 1960s that the company needed to add upholstered furniture (living room, family room and accent seating) to round out the Bassett assortment of products. This had been accomplished in 1963 with the purchase of Prestige Furniture Company, an upholstered furniture manufacturer located in Newton, North Carolina. The purchase of Prestige was good timing for me because I was also able to spend time in the Newton plant learning about the manufacturing of upholstered furniture. There, I got to see the cutting and preparation of the frame parts, and the assembly of those parts into wood frames for upholstered sofas and chairs.

Next came the spring-up area, where the springs were attached to the frames to provide the right amount of comfort in the backs and seats of the products. I also got to see the cutting and sewing of the decorative fabric covers for the products, as well as the upholsterers applying the covers to the upholstered product frames and neatly placing the seat and back cushions inside the sewn fabric covers. I was amazed at the skill of the fabric cutters, sewers and upholsterers working in the plant.

My time in the Bassett wood and upholstery furniture plants gave me a good appreciation for what is involved in the manufacturing of home furnishings products. In addition to learning about manufacturing, I experienced something I had not anticipated, and that was the obvious pride that the factory workers showed as they explained to me what was involved in their individual jobs. In the wood plant, I was especially impressed with the care the workers took in applying the finishing materials so that the overall finish was very smooth, with no runs or missed spots to interfere with the overall appearance of the piece. In the upholstery plant, the sewers took time to explain how they made sure the seams were straight and the fabric designs were perfectly matched.

MARKETING SIDE OF THE BUSINESS

As the person who coordinated my schedule at Bassett, Mr. Meadors made sure that I was exposed to the entire marketing cycle of the Bassett Furniture products. The timing of my leave of absence from High Point College allowed me to see most of the planning leading up to the April 1979 furniture market.

Soon after the first of the year, I sat in on some of the product planning meetings, which were conducted by Charlie Bassett, vice-president for product development. These meetings were held to determine which current products were selling profitably and which should be replaced by new introductions at the April furniture market. Ideas were solicited from key Bassett sales representatives in selected areas of the United States. These sales representatives were important to the product planning process because of the insight they had gained from being out in their sales territories, learning what products were selling and what voids they saw on retail sales floors that could be filled with Bassett products.

Because of its large size and its volume of product introductions, Bassett had furniture designers on staff who drew sketches of product ideas presented at the meetings. (This was a different process than the one used by most smaller manufacturers. These companies contracted with independent freelance designers to do their design work.) The most promising sketches were then converted into mechanical drawings. The next step was for the sample departments in the plants to construct mockups or samples of the products from these mechanical drawings. Key dealers were invited to review these samples during a time referred to as "premarket," which was approximately one month prior to the April furniture market. If a dealer committed to buying a sufficient quantity of a certain product during premarket, that product would be an "exclusive" for that dealer and would not be shown to others at the furniture market. All dealers would then be invited to shop for the remaining products at the furniture market.

Learning about the product development cycle gave me an appreciation for the furniture industry as a fashion industry. New products were developed with careful consideration given to what were perceived as emerging fashion trends in home furnishings by Bassett's targeted customers. For example, informal country-style products seemed to be preferred by Bassett's customers in the late 1970s.

Along with consideration of the products as home fashions, thought was given to developing a marketing story for various pieces and collections. This, to me, seemed particularly challenging because as the world's largest furniture manufacturer, Bassett produced most categories of home furnishings, such as dining room, adult bedroom, youth bedroom, home office, occasional tables and baby room furniture (often called juvenile furniture by some professionals in the industry), which included such items as cribs and changing tables, among many others.

An interesting marketing story for baby room furniture was to develop collections that could be advertised as "crib to college." In other words, in some juvenile furniture collections, they avoided using bright colors, finishes or designs that would clearly label the furniture as "little boy" or "little girl" furniture. The use of more traditional finishes and designing the products so they could be used by older children and teenagers appealed to parents and grandparents who didn't want to buy furnishings that would have to be replaced in a very short period of time. Of course, other considerations, such as safety regulations for baby cribs and flammability standards for fabrics and other materials, also had to be taken into account.

Another aspect of furniture marketing I was exposed to was the job of the manufacturer's representative. I was invited to ride with Bassett's North Carolina representative as he drove through his territory, calling on retail stores that sold Bassett furniture. In most cases, he had called the store ahead

Typical Bassett Furniture products of the 1960s and 1970s. *Courtesy of Bassett Furniture Industries.*

of time and made appointments with the buyer or owner. It was obvious that he was well aware of what each retail store had in stock and on order before we arrived. On arriving at the store, his first stop was to go to the warehouse to see if there were any quality issues that needed to be taken care of. This visit to the warehouse allowed him to know the retailer's situation in regards to any quality issues or damages to the dealer's Bassett products before the time of his appointment. It was clear that he respected the retailer's time and was trying to be a true partner so that it would be a win-win situation for both Bassett and the retail store.

Spending time learning the manufacturing, product development and marketing sides of the business by seeing the Bassett operation was invaluable as I began preparing for my classes at High Point College and later writing a furniture marketing textbook.

Mr. Meadors recently told me that they enjoyed having me spend time with the management and employees at Bassett. I could tell that they really enjoyed "showing me the ropes." They were proud of the various aspects of their business and were eager to share their experience and knowledge with me. The management of Bassett Furniture seemed to have a high regard for me and the partnership that High Point College was developing with the furniture industry. They hoped that this partnership would lead to a supply of outstanding future employees for the industry.

MY FIRST VISIT TO MARKET

My first visit to the High Point Furniture Market was in the spring of 1979. I was eager to go to the furniture market because I had driven by the market buildings since moving to High Point in 1974 but had never been inside. I had been told that the market was "to the trade only," except for a few invited guests, and I felt privileged because I was probably one of the first invited guests from High Point College to actually spend time exploring the market.

My first market experience began in the large freestanding Bassett Furniture Showroom on Business I-85 on the edge of Thomasville (a relatively small town just outside the city of High Point). It was not unusual in the late 1970s for furniture manufacturers to have freestanding showrooms outside of downtown High Point or even have their showrooms next to their factories, many of which were located a considerable distance from High Point in North Carolina. These were towns known for the manufacturing of furniture, like Lexington, Lenoir and Hickory. At that time, the Hickory Furniture Mart building in Hickory, North Carolina, was a wholesale market showroom building for several of the manufacturers in the western part of the state.

Bassett Furniture's freestanding market showroom building. *Courtesy of Bassett Furniture Industries.*

BASSETT PREPARES FOR MARKET

In the weeks before market, it was obvious that the furniture market was very important to Bassett, as it was to the other companies in the furniture industry. Most of the talk I heard involved making sure that everything was in order so that the market visitors would have a good experience. Above all, it was hoped that they would place orders for Bassett furniture.

THE SHOWROOM WAS READY

The showroom designers had spent a lot of time and effort making sure that each part of the showroom was clean, newly painted and attractive for the buyers. The market samples of the wood furniture were checked to make sure there were no scrapes or other damages to their finish and that each had the proper decorative hardware. Upholstered furniture samples were also checked to determine that the upholstering had been done properly and that the new fabrics were being shown. The samples were then placed in their proper positions on the showroom floor, and the market set up team made sure that everything was ready for the first day of market. The scene struck me as similar to getting ready for the opening night of a theatrical production.

The Sales Representatives Were Ready

In 1979, the largest percentage of Bassett Furniture's sales seemed to be through large retailers, such as Levitz, Sears and JCPenney, but they also had a large variety of other dealers selling their merchandise throughout the United States and internationally. The Bassett sales force was comprised of approximately 350 sales representatives from all over the country. It was a learning experience for me to be allowed to attend the sales meetings that took place just prior to the market. This was the time when all the sales representatives came together to learn about the products that were being introduced at the market and any new merchandising programs that were available.

After the meetings, I was amazed as I watched the sales representatives learning all they could about the new product introductions before their customers came through the front door. (It was not unlike High Point College students cramming for final exams.) I discovered that each salesperson worked with the dealers from their territories. When the doors opened, I watched from a distance as the sales reps met the dealers and made professional, effective presentations of the products they felt would be best for each dealer's store.

I also learned from the sales reps that a key time for them was what many in the industry called "post-market." This was the period immediately after the market when they visited the dealers in their stores. During this visit, they followed up on what they had talked about with each dealer at market and "closed the deal," as this was often when the most orders were actually written.

Downtown Showrooms: Each Designed to Be Different from the Competition

Following my orientation to the market at Bassett, I made numerous visits to the market showroom buildings and freestanding showrooms in downtown High Point. As I searched for a place to park and began walking toward the showrooms, it became more and more apparent to me that the central part of the city was dominated by the furniture market. It was then that I realized for the first time how big the furniture market really is.

SAWING THROUGH -- Harold Rodenhouse [left], president of Hekman Furniture, puts aside one part of the six-foot wooden bar placed across the front doors of Hekman's new showroom at 200 N. Hamilton in High Point. He and Richard Shwayder [right], president of the home furnishings division of Beatrice Foods, sawed through the bar in the official grand opening festivities of the new showroom.

33

FURNITURE/TODAY, April 21, 1979

The grand opening of Hekman Furniture's market showroom, circa 1979. *Courtesy of Furniture Today.*

For the most part, the downtown furniture market showroom district had grown up on either side of Main Street, approximately in the center of the city of High Point. It was interesting to learn that although the largest percentage of showrooms were in the Southern Furniture Exhibition Building, many other companies chose to have their showrooms in smaller showroom buildings or in freestanding buildings. Most of these are on the streets close to the big showroom buildings with their front doors opening directly onto the street, much like retail stores in downtown areas of large cities. But regardless of their location, each showroom was deliberately designed to enhance the appearance of the products and stand out from the competition.

A majority of these showrooms were very open and informative after I showed them my market badge and explained why I was there. I was amazed at the many different types of showrooms I visited. Most of the exhibitors seemed to be making a fashion statement. They were either focusing on one or a limited number of styles. The smaller companies were usually showing products within one or two product categories, while some of the larger companies were showing furnishings for most of the different rooms of a house. Regardless of the size and product category, the market showroom displays seemed to be saying to the buyers, "We have the furnishings to appeal to the customers who are coming into your stores this year."

Although there were obvious differences in the merchandising approaches of their market displays, most exhibitors were trying to make their products as attractive as possible to the retail store buyers. Some were showing their products in an easy-to-compare arrangement. For example, a company that manufactured occasional tables (usually correlated sets of a sofa table, two end/lamp tables and a coffee/cocktail table) showed all of their products on risers around a large room. Each correlated table group was shown together so the buyer could easily choose from the light wood, dark wood and painted tables from whatever style assortment they were featuring that market.

Some exhibitors were showing their products in room settings that might give the potential buyers ideas of how the products could be most effectively displayed in their retail stores. Others might be showing new or different types of products. For example, La-Z-Boy Inc. might be showing products in room settings in one part of the showroom, and in other parts of the showroom, it might be showing a new type of product, such as "close to the wall" recliners.

Other exhibitors were showing a merchandising program to the dealers. For example, one manufacturer of promotional upholstery was showing an L-shaped sectional sofa with the L composed of chair-width sections, each

Products shown in La-Z-Boy's market showroom. *Courtesy of La-Z-Boy Inc.*

upholstered in a different fabric to show the seven or eight different fabric choices they were offering at that market. The idea was that the dealers could put this "demonstration" sectional in their stores so that the retail customers could easily see their options and choose which fabric was best for them.

A more common merchandising technique in upholstered furniture seemed to be to offer a cutaway chair or other upholstered piece so that retail customers could see the inner construction of their products, combined with a wall display of the available selection of fabric swatches for consumers to choose from while they were visiting the store. The idea was for the retailers to put this point-of-purchase display in their stores to simplify the consumer's buying decisions. They could see the quality of the product and choose the right fabric for their home.

There was one company with a showroom in the Southern Furniture Exhibition Building that I remember went to a considerable effort to entice buyers to step into their showroom. It was a rather long showroom with large windows, so market visitors could look into the showroom before actually entering. The company mostly sold brass accessories, such as statues and vases, and lighting, such as table lamps, floor lamps and chandeliers. As I looked through the windows and into the showroom, I saw a number of female models, all clad in brass-colored hotpants outfits, walking around and smiling as if to say "come on in." It was obvious that the company assumed the buyers who were coming to see their products were mostly men. It reminded me of photographs I had seen of the new car introductory shows in Detroit, which also used attractive female models. In all my visits to market since then, I have never seen a showroom quite like that one.

I do remember a couple of other companies that went to considerable efforts to increase traffic or make buyers feel welcome to their showroom. A showroom that I passed by a few times always had a very pretty blond lady with a broad smile working at the front desk to greet the buyers. On one of my trips by this showroom, I thought I was seeing double. There was what appeared to be the exact same pretty blond lady standing behind the receptionist. I learned that they were identical twins from New York who had been brought down to work at the furniture market.

Many companies also provided a complimentary lunch for the buyers. One in particular featured lunch with a different entrée each day. One day could be lobster flown in from New York cooked a particular way; the next day would be something different, but it was always good.

OTHER MARKET ACTIVITIES
BESIDES SHOPPING SHOWROOMS

I also discovered that activities other than retail store buyers shopping showrooms for merchandise take place at market. This seemed to be especially popular with the smaller- and medium-sized retailers. While their main objective was to shop for furniture for their stores, there were also a variety of educational seminars that these retailers and other market visitors could attend on various subjects ranging from retail store design to how to better manage their store's finances.

Retailers also visited the National Home Furnishings Association Retail Resource Center. Here, retailers could learn about a variety of services

Market personnel distributing information about products and events. *Courtesy of International Market Centers.*

that could help make their businesses more successful. For example, one vendor might offer a retail advertising program, complete with sample advertisements and point-of-purchase materials for retail furniture stores. Another might have a retail sales training program to help increase the professionalism and productivity of retail floor salespeople, while still another might offer warehousing programs to make the retailer's operations more efficient.

As I reflected on my first visit to the furniture market, I realized how important the furniture market is, and for the first time, I understood the critical role it plays in the distribution of home furnishings products.

Chapter 4

SPECIALIZED FURNITURE EDUCATION COMES TO CAMPUS, AND WE START SPREADING THE WORD

Beginning with the fall semester of 1979, the bachelor of science degree in furniture marketing was approved as a major within the Earl N. Phillips School of Business at High Point College. The major was based on the in-depth industry research I had conducted during my leave of absence, along with the advice and counsel of Dick Burow, the retired president of Chicago-area Kroehler Furniture Company, who had been retained as a consultant during the initial implementation of the program. An intentional effort was made to use the local resources of the industry as much as possible for field trips, guest speakers and internships.

BACHELOR OF SCIENCE DEGREE IN FURNITURE MARKETING

The bachelor of science degree in furniture marketing (later changed to the bachelor of science degree in home furnishings marketing to be more inclusive of all home furnishings products) offered courses that were helpful to students preparing for a meaningful career in the home furnishings industry. Located in the School of Business, the major also included a sizable number of business courses, such as Principles of Marketing, Business Communications, Consumer Behavior and Business Ethics, in addition to industry-specific courses.

The industry-specific courses included Introduction to Furniture, Furniture Retailing, Fundamentals of Interior Design, Furniture Marketing/Manufacturing, Furniture Sales Development and a senior seminar in Home Furnishings Marketing Strategy. The first course offered in the fall semester of 1979, Introduction to Furniture, was based largely on my experience seeing the products made in the Bassett case goods and upholstered furniture factories, my observations at the furniture market and input from Burow. This course was designed to give students an understanding of insider terminology, the products sold in the industry and the basics of how the industry works.

Other helpful courses that covered the inner workings of the industry were phased in over the next couple of semesters. Introduction to Interior Design was implemented to give the students an appreciation of the principles of interior design and how furniture, as a functional art form, fits into the overall design of a space. My Bassett experience was useful again, as I developed additional courses in areas such as the marketing, merchandising and sale of home furnishings products.

Home Furnishings Marketing

The important function of making modern man comfortable and aesthetically satisfied in his home and his world—that's what home furnishings marketing is all about. To put it another way, it is concerned with creating harmony between people and their homes. It's an infinitely varied, satisfying, and creative field, and it's a particular specialty of the High Point College curriculum.

HIGH POINT COLLEGE
HIGH POINT, NORTH CAROLINA

A brochure containing the philosophy of the home furnishings marketing program. *Courtesy of High Point University.*

A number of rather specific areas were covered in the curriculum. For example, channels of distribution were covered so students would understand the necessity of having the right channel for the type of home furnishings products being marketed to carefully targeted customers. The curriculum also included the factors involved in merchandising products at both the manufacturer/wholesaler and retail store levels. Among these factors are product design, pricing and promotion, as well as consumer motivation and buying behavior. Essentially, they were designed to cover the factors that would likely cause a consumer to purchase a particular product.

Wherever possible, class projects were assigned so that students could understand as many of the industry-related marketing variables as possible. For example, the Furniture Retailing class required a group project in which students were placed on store teams of approximately five students each.

Each team was given a price range of merchandise for the store (low price, medium price or high price). The students were then to work together on the following:

1. Studying the demographics of the population in various geographic areas so that they could choose an area where a store in their assigned price range might be successful.
2. Choosing an assortment of merchandise for the major home furnishings categories from vendors with products in their assigned price range.
3. Drawing a floor plan of the store indicating the placement of the products on the sales floor.
4. Developing a sales and advertising plan, including number of salespeople, types of advertising and the media that would be most effective in presenting the advertising.
5. Formulating a warehousing plan and plan for delivering the products to customers.
6. Describing any credit plans that could be offered so customers could have the option of financing their purchase over a designated period of time.

A senior capstone seminar course was required for all furniture marketing majors. This course was designed to allow students to understand key marketing areas faced by professionals on both the manufacturing and retail levels. In addition to class presentations by the instructor and selected guest speakers, students were given ten different case studies to read and asked to answer questions regarding the situation presented in each case. A different case was assigned each week for ten of the fourteen weeks of the class, with the requirement being that students had to write up their answers to the questions and be prepared to participate in class discussions of the case. The objective of the course was to build the students' decision-making skills so they would have the confidence to successfully fill marketing-related positions in the industry.

In addition to the cases, the students were assigned to work in teams to research a current topic of interest to the home furnishings industry and work together to prepare a written report that they would jointly present in class, using visual aids to help explain the project. Professionals from the industry were invited to sit in on the oral presentations and rate the students on both the information presented and their presentation, including their ability to get the message across and their professionalism (such as how they were dressed).

An example of one topic assigned was "Nontraditional Channels of Distribution for Home Furnishings." Under this topic, one team might be assigned to study online shopping (e-commerce) as a viable avenue for marketing home furnishings. They were to look at whether the channel was increasing or decreasing and the long-term advantages and disadvantages of computer shopping for furniture. They were to assume the professionals visiting the class were members of a corporate board of directors considering either increasing or decreasing their efforts to sell home furnishings online. Their goal was to provide valuable information to aid the company in making its decision.

ADVISORY BOARD OF INDUSTRY EXPERTS

An advisory board of industry experts was put in place to not only help in the initial development phases of the major but also to meet semiannually to provide ongoing advice and counsel regarding the changing needs of the industry. The intent of the board was to plan an educational program that would continue to be relevant on an ongoing basis. The board was to also assist in public relations to spread the word about the furniture marketing program in an effort to increase student enrollment; and look for sources of funding for scholarships and other needed assistance for the program.

The advisory board chairman was to be someone from the industry, and the members of the board were to be representative of as many different segments of the industry as possible. The following list of some of the early board members gives a snapshot of the intentional diversity of the board:

- Clarence Smith of retailer Haverty Furniture, an early advisory board chair.
- Joe Carroll, publisher of industry trade publication *Furniture Today* (he was also a later advisory board chair).
- George Erath of industry supplier Erath Veneer Co. (He had also served as chairman of the High Point University Board of Trustees and been awarded an honorary doctorate by the university.)
- Joe Meadors of manufacturer Bassett Furniture.

- Tom Muzekari of industry supplier Quaker Fabrics.
- Joanna Maitland, director of communications for UFAC (Upholstered Furniture Action Council) and the former director of consumer affairs for the Sperry and Hutchinson Company.
- Ben Willis of retailer Willis Wayside.
- Pat Plaxico, prominent High Point Interior Designer.
- Frank Hanshaw Jr. of wholesaler Huntington Wholesale Furniture Co.

Informational Booth at the International Woodworking Fair (IWF)

For several years, High Point University was one of a number of colleges and universities that was provided a booth at the International Woodworking Fair (IWF), a trade show for the furniture manufacturing, architectural woodwork and custom and general woodworking industries. This allowed a representative of the university to distribute information on HPU's home furnishings–related programs. The IWF, which, for several years, was held at the Georgia World Congress Center in Atlanta, provided HPU exposure to owners and managers of wood furniture manufacturers throughout the United States. This booth was a very important way of increasing the awareness of the university's home furnishings–related majors to the furniture industry, especially during the 1980s, when the majority of wood furniture for the U.S. market was still being manufactured in the United States.

Unique Teaching and Learning Opportunities Due to Being Located in High Point

From the very beginning, field trips to local businesses have been integral parts of the home furnishings courses at High Point University. In the fall semester of 1979, students obtained product knowledge through field trips to companies like Thomasville Furniture Industries, where they were exposed to wood furniture manufacturing; Pearson Furniture Company, where they were exposed to upholstered furniture manufacturing; and Hayworth Roll and Panel, where they learned about veneers and plywood

Students learning about upholstered furniture construction on a local fieldtrip. *Courtesy of High Point University.*

Joe Carroll, the publisher of *Furniture Today*, speaking to home furnishings students. *Courtesy of High Point University.*

panel construction. Although these factories eventually closed, they have been replaced by field trips to other furniture factories within an easy driving distance of the campus so that students can continue to learn about various types of furniture manufacturing.

Sales, merchandising and marketing were presented through field trips to retail stores and showrooms in the semiannual furniture market, as well as invited guest speakers to the classroom and having the students work together on case studies. Guest speakers who have visited the classes and presented much-needed information have included such individuals as CEOs and other managers of home furnishings manufacturing companies, representatives of industry trade publications, various management officials of retail stores, furniture designers and representatives from industry trade associations and suppliers to the industry. These field trips, guest speakers and case studies all brought realism into the various furnishings-related courses. Students were also encouraged to participate in internships and other part-time work experiences, such as working for local home furnishings companies and helping staff the High Point Furniture Market showrooms to extend the learning process beyond the classroom.

OUT-OF-TOWN INTERNSHIPS

Although most of the internships were with local companies in the High Point area, the university, for a number of years, had a program in which a current High Point University student could do a summer internship in another part of the country, provided a faculty member physically visited the intern at their work location. The expenses of the visit had to be paid for by the intern or their family. Two of my experiences supervising summer interns made for some interesting memories—one with me flying across the state of North Carolina in a private plane, and for the other, I flew halfway across the United States to see a very modern, up-to-date manufacturing operation and the most immense furniture warehouse imaginable.

The first internship was at Murrow Furniture Galleries in Wilmington, North Carolina. The store manager's son was a student in the High Point Furniture Marketing Program, and he wanted to do a summer internship in the Wilmington store. Murrow Furniture Galleries was owned by Coolidge Murrow, who, at that time, also owned Country Furniture, a retail furniture store in High Point. Murrow was a pilot and had a private airplane that he flew between the Piedmont Triad International Airport near High Point and the Wilmington Airport. He arranged for me to fly with him in his plane so I could visit the High Point University student interning at his store. I not only got to see what the student was doing at the store, but I also had a very enjoyable, scenic trip across North Carolina on a beautiful, sunny day.

A second memorable internship was at the Ashley Furniture headquarters and plant in Arcadia, Wisconsin. The son of one of the vice-presidents of Ashley Furniture Industries Inc. was a student in the High Point University Furniture Marketing Program. He wanted to do a summer internship at the Ashley headquarters in Arcadia. Since an HPU faculty member had to visit the internship site in order for him to get academic credit for his internship, Ashley Furniture Company flew me to Arcadia so I could observe him. This trip gave me the opportunity to both observe what the student was doing and see the Ashley Furniture operation.

One particularly new and different type of production for me was to see was the manufacturing of furniture with a polyester finish, which was popular at that time. (It was the Ashley Millennium line, which featured a shiny, colorful finish requiring the production environment to be very clean and dust-free.) I also got to see the largest furniture warehouse I had ever seen; some of the managers and other employees even rode bicycles to get around inside the building. It was an efficient warehouse, where imported products were sorted and stored before being shipped out to Ashley Furniture stores.

These are only two of the many internships that allowed me to make sure students were getting valuable real-world experiences and that helped keep me current as to what was happening in the industry.

First Furniture Marketing Majors Graduated in 1981

In 1981, the first High Point students graduated with the bachelor of science degree in furniture marketing. Many of these graduates were High Point College students who had entered the program at the beginning of their junior year or High Point area residents who had transferred in from other colleges or universities when they found out about the program. Recently, I talked with one of the graduates of the class of 1981, and he provided an interesting perspective on how his educational experience at High Point

John Gallman, a 1981 graduate and manufacturer's sales representative, showing a Miles Talbott chair. *Courtesy of Richard Bennington.*

University impacted his choice to pursue a home furnishings–related career and his subsequent success.

The 1981 graduate I talked with was John Gallman. In 1979, John had just graduated with a two-year degree from Davidson County Community College, located just a few miles outside of High Point. He was entering High Point College the same year we were starting the furniture marketing program. John came to High Point College to pursue a degree in business administration. In talking with him as his advisor, I asked him if he would like to take some courses in the new furniture marketing major. The idea intrigued him, so he enrolled in Introduction to Furniture during his first semester at High Point College.

John liked the furniture marketing courses more than the basic business courses, so he changed his major and completed the furniture marketing major in time to graduate in the spring of 1981. After graduation, he took a job in sales with Woodmark Originals, a high-end upholstered furniture manufacturer in High Point. Regardless of the fact that this was his first job out of college, he was employed as one of four manufacturer's representatives covering the entire United States. He liked working for Woodmark Originals and stayed with them several years, eventually becoming vice-president of sales and marketing. When the company was sold, John became an independent, multiline sales representative, a job he still holds today.

When asked what skills he thought would be most valuable to today's graduates, he said, "First, people skills and second, computer skills." Graduating with a minor in psychology, along with his furniture marketing major, gave him the perspective that graduates need well-developed people skills. He believes that graduates should develop what he called "chameleon skills," which involves the ability to communicate with a variety of people on their own level. This relates well to his sales representative job, which involves making the sales presentation that is best for each individual customer.

He also believes that to be successful in today's business world, students should develop good computer skills, especially the ability to work with Excel spreadsheets. In other words, students should understand the types of data that are needed in whatever job they have and should be able to effectively work with it.

1985 BRINGS THE FIRST
FURNITURE MARKETING TEXTBOOK

In 1985, Fairchild Publications in New York published the first edition of my furniture marketing textbook, *Furniture Marketing, From Product Development to Distribution*. This textbook was the product of all my cumulative experiences in the furniture industry—from my six months with Bassett Furniture and numerous visits to the High Point Furniture Market and some of the other U.S. furniture markets to visits with selected retailers throughout the United States.

The book was written to be a basic resource guide to home furnishings products, the home furnishings industry and the necessary factors that must be considered in successfully marketing home furnishings products in the United States. It was designed to be used either as a textbook in college or university classes or by burgeoning professionals as an orientation to the industry. It explains what the various product categories are, the basics of developing products with consumer-attracting attributes and the variables involved in selling home furnishings products. A second edition of the furniture marketing book was published in 2004.

The first edition of the Furniture Marketing textbook that was published in 1985.
Courtesy of Richard Bennington.

BSBA (BACHELOR OF SCIENCE IN BUSINESS ADMINISTRATION) IN SALES: CONCENTRATION IN THE FURNITURE INDUSTRY

Professor Larry Quinn, the director of the High Point University Sales Education Center. *Courtesy of High Point University.*

The bachelor of science in business administration in sales with a concentration in the furniture industry has been instituted in the Earl N. Phillips School of Business for students who are contemplating a career in sales in the furniture industry, either at the retail level, the wholesale level or the global supply chain level. The mission of the BSBA in sales is to provide students with the skills and experience necessary to significantly improve their value to the selling profession, showcase the talent of High Point University students and give employers a source from which they can recruit motivated, qualified new sales talent.

The courses required for the program are designed to provide students with a solid background of business classes, a basic knowledge of residential and commercial furnishings, an understanding of the interior furnishings industry and the knowledge of sales as well as basic selling skills. Specific selling courses that are offered include Sales in a Dynamic Environment, Negotiations of Retail Selling and Sales Leadership. Selling skills are taught in the state-of-the-art Harris Sales Education Center, located in Cottrell Hall on the High Point University campus.

Chapter 5

INTERIOR DESIGN IS
ADDED TO THE MIX

In the years after 1981, when the first furniture marketing majors graduated and began working in the industry, a variety of other students began asking about the program. Part of the appeal of the forty-five-semester-hour major in home furnishings marketing was that one of the home furnishings–related courses was Principles of Interior Design. During the mid- to late 1980s, current and prospective students increasingly asked the High Point College faculty to offer additional classes in interior design. The most sizable group asking for these classes were women who were interested in possible careers in the industry but whose interests seemed to be more in-line with interior design. They liked the idea of working with home furnishings products but really wanted to know more about creating retail store or market showroom spaces to enhance the sale of these products. They also wanted to learn how to use furnishings as tools to help enhance the efficiency and comfort of living or working spaces.

A Minor in Interior Design
Sounds Like a Good Place to Start

In 1991, the same year that the name of High Point College was changed to High Point University, Dr. Elizabeth Dull was hired as the first full-time interior design instructor, and a minor in interior design was added to the curriculum. At first, most of the students in the interior design minor were furniture marketing majors, although the minor was open to other High Point College students. The minor contained classes like History of Interiors and Furnishings, Interior Design Principles, Textiles, Visual Presentation, Lighting Design and Commercial Space Planning for Stores and Showrooms.

Our Interest Is Increasing; Let's Add a Major in Interior Design

Beginning with the 1994–95 academic year, the bachelor of science in interior design was added to the curriculum. The new program was designed to complement the major that was already available in home furnishings marketing. The first students to complete the major in interior design graduated in the spring of 1996.

Administratively, it was decided that the development of the interior design major within the Earl N. Phillips School of Business would be a unique feature and would allow the interior design and home furnishings marketing majors to share resources, such as faculty, industry-specific library materials, selected courses and designated scholarship funds. The complementary nature of the two majors was seen as providing the widest range of opportunities for students interested in careers in the furniture industry. The core classes of Introduction to Furniture, Furniture Marketing, Furniture Manufacturing, Furniture Retailing and Introduction to Interior Design were required for both majors, which was seen as desirable by both furniture industry leaders and interior design professionals.

Wherever possible, the interior design and home furnishings programs worked together so they could learn from each other. For example, on the retail store team project, described earlier in chapter 4, the interior design students selected the merchandise and drew out the plan for the retail store sales floor. At the same time, the home furnishings marketing majors worked on other parts of the project, like selecting the specific store location and formulating the sales and marketing plan, warehousing/delivery plan and the consumer credit options for the store. This reinforced the often-voiced opinion by professional interior designers that an understanding of manufacturing and retail operations, as well as a thorough understanding of furniture construction, is a definite plus for the interior designer. Another option for interior design majors looking to broaden their knowledge was to

An interior design fieldtrip to the High Point Furniture Market. *Courtesy of Cathy Nowicki.*

An interior design fieldtrip to Biltmore House in Asheville, North Carolina. *Courtesy of Cathy Nowicki.*

also minor in home furnishings. Graduates who majored in interior design with minors in home furnishings marketing have been very successful in finding employment in their chosen field.

Field trips, guest speakers and internships were considered invaluable to the interior design major, just as they had proven to be in the home furnishings marketing major. Interior design fieldtrips are conducted to both destinations in the local High Point, Greensboro and Winston-Salem areas and to destinations farther away, usually within a day's drive of the campus. In this example (the image on page 65), the interior design students are learning about wholesale showroom market design by visiting showrooms at the High Point Furniture Market. Students also go on fieldtrips outside the High Point area. For example, students recently traveled to Asheville, North Carolina, where their perception of interior design was broadened by visiting Biltmore House, the largest house in the United States.

INTERIOR DESIGN GRADUATES ARE USING THEIR TRAINING TO ACHIEVE SUCCESS IN THE MARKETPLACE

Sarah Bennington Hogan, for example, came to High Point University to major in interior design. When she graduated from the university in 1999, she began her career by working as a visual merchandiser for Havertys Furniture in their Winston-Salem, North Carolina store. After gaining experience with Havertys, she took the position of visual merchandiser with Boyles Furniture, at that time, a relatively large medium- to high-end retail store in her hometown of High Point. When the Boyles store closed, she worked for manufacturer French Heritage as showroom manager for their High Point Market showroom. She also traveled to retailers around the country, where she worked with furniture stores to design their French Heritage galleries and displays.

Ultimately, Sarah left French Heritage for employment with Kreber Inc., a full-service photography studio in High Point, where she worked as a photography set designer. She does set design for both still photography and video. Her sets were used in photographs for catalogs, print advertising and websites for manufacturers and resellers—both traditional brick-and-mortar and e-commerce retailers. She liked her job, especially because every day brings something different.

HPU interior design graduate Sarah Bennington Hogan working as a photography set designer. *Courtesy of Kreber Inc.*

After nine years with Kreber Inc., Sarah decided to use her design talent and knowledge in a different way by accepting the position of visual merchandiser with Aspenhome, an Arizona-based manufacturer and importer of medium-priced wood furniture for the home. Her job is to plan and execute the merchandise strategy for the company's High Point and Las Vegas market showrooms. In each location, she creates visual displays to showcase the furniture and build brand integrity.

In her education at High Point University, she feels that exposure to the furniture industry helped immensely in bringing her success. Field trips to retailers and manufacturers helped her gain product knowledge but also helped her understand the entire merchandising process. She has found that, regardless of her area of employment, knowledge of the furniture industry as a whole and how each part works together has been an essential part of her success.

Sarah thinks that current students need to learn, if possible, through real-life internships. As she put it: "At High Point University, I learned the tools I would need, but once I got out in the real world, I learned how to use those tools effectively to get the job done."

LET'S GO FOR GOLD: CIDA ACCREDITATION

In 2007, the interior design major was accredited with a three-year interim visit by the Council for Interior Design Education (CIDA), the widely accepted accrediting organization for interior design programs at colleges and universities in the United States and internationally. CIDA has identified and developed and is engaged in promoting quality standards for the education of entry-level interior designers. Institutions seeking accreditation must write an in-depth report, carefully detailing all aspects of their interior design program. After reading the report, a visiting team from CIDA makes a site visit to determine whether or not the institution meets their quality standards. Some say that satisfying the CIDA standards and obtaining accreditation from that organization is like meeting the "gold standard."

Following the interim visit in 2010, the High Point University interior design major was fully accredited by CIDA. It was then reaccredited in 2015, following the same examination process that was conducted in 2007 and 2010.

High Point University's certificate of accreditation from Council for Interior Design Accreditation (CIDA). *Courtesy of Richard Bennington.*

Another advantage of being a CIDA-accredited program is that students graduating with a major in interior design from HPU have all the education requirements needed to sit for the National Council for Interior Design Qualification (NCIDQ) examination. The NCIDQ is the only nationally recognized professional competency exam for interior designers in the United States and Canada.

Interior Design Graduate William "Bill" Lyon

William "Bill" Lyon, a native of Creedmore, North Carolina, decided to attend High Point University for two reasons: first, it had an accredited interior design program, and second, it was close to the High Point Furniture Market. When he graduated in 2013 as an interior design major, he had his employment already in place because of his experience in working at the High Point Market. Beginning in the fall semester of his sophomore year, he worked with Lillian August, who has a higher-priced, design-oriented showroom in the Hamilton-Wrenn showroom district of the High Point Market. Bill's job for the first market was to help the interior designers (who are the bulk of the visitors to the Lillian August Showroom) with fabric samples and other needed information. He was so well liked by Lillian August and her staff that he continued working with them until he graduated—a total of six markets.

Bill was so successful with his market job that by the time he graduated, he was walking the designers through Hickory White and other showrooms that were associated with Lillian August. He also got to the point where he would give the visitors advice on where they should go to shop for accessories or other compatible products. When Bill was nearing graduation, Lillian August invited him to join her as a full-time employee in Connecticut. He accepted her offer of employment and advanced through a number of positions until he reached the position of senior designer at her store in Greenwich, Connecticut.

Bill loves his career, and, to borrow a few phrases from his biography as printed on his website, his mission is to make "the world a little more stylish every day." He "brings to his clients a love of fashion and furnishings carefully integrated with a deep understanding of the international design world, both honed to a razor's edge through education and extensive industry experience" (www.williamlyondesigns.com/aboutwilliam). Among

Above: Interior design graduate William Lyon working with Lillian August in the Lillian August Market Showroom. *Courtesy of Cathy Nowicki.*

Left: Interior design graduate William Lyon working with a customer as a professional designer. *Courtesy of William Lyon.*

his accomplishments are assisting with the opening of the Lillian August Atelier store in Greenwich, Connecticut, designing the box office for both the 2016 and 2018 Greenwich International Film Festivals and having one of his residential designs featured in *atHome* magazine.

He also excelled academically at High Point University and received the Celia Moh Scholarship for his senior year. When asked about what from his HPU courses was most important to his success in his career, he thought the factory tours, class projects and space-planning skills he learned at HPU were all invaluable. The skills he thought students should learn in their classes at HPU are things like time management and how to "read" the client so that the designer can create a design just for them. Learning how to dress appropriately is also important, a topic included when HPU teaches "life skills" needed by graduates.

After six years with the Lillian August organization, he left that company and started his own design firm, William Lyon Designs. There, he is able to use both the knowledge he acquired at High Point University and the real-world experience he gained through working with Lillian August to tailor his designs to an exclusive United States or international clientele.

Visual Merchandising Design: A Logical Addition

In 2010, High Point University decided to move the Department of Home Furnishings and Interior Design out of the Earl N. Phillips School and combine it with the majors in studio art and graphic design to form a new School of Art and Design. Following the establishment of the School of Art and Design, a new bachelor of science degree in visual merchandising design (VMD) was created. A minor was also developed that was to be available to students majoring in other areas. These new programs were added to the curriculum in 2013, with the first visual merchandising design majors graduating in 2016.

The visual merchandising design program was established to train highly creative students to be able to combine business knowledge with their creativity to stimulate a positive consumer response, thus increasing the likelihood that consumers will buy a particular product. This could be achieved through displays in retail stores, on websites, in market showrooms, in advertising photography or anywhere else that an eye-catching, carefully

A store display done by High Point visual merchandising design students. *Courtesy of Benita VanWinkle.*

targeted visual message is helpful in selling a product. The training provided by the visual merchandising design majors or minors would equip graduates for jobs in areas such as visual display design or management, product line development, retail store design, market showroom design, exhibit and display design, photography set design, merchandise management, retail buying, fashion product merchandising and visual marketing.

High Point University's visual merchandising majors have the unique opportunity to not only graduate with a four-year degree, but also to study abroad during the spring semester of their junior year at the Lorenzo de' Medici Institute in Florence, Italy, if they so choose. Because of the unique configuration of the major, they can get their degree and study for a semester in Florence without extending their time at High Point University beyond four years.

The idea that visual merchandising design was a good idea has been proven by students like Addison "Addie" Gantt. Addie came to High Point University from Salisbury, North Carolina, thinking about majoring in history. That all changed when she found out about the visual merchandising design major. She decided on VMD as her major and began actively preparing for a career in the home furnishings industry while she was a student at High Point University. She worked for Klaussner Furniture, helping them prepare their showroom for every High Point Market from her freshman

High Point University visual merchandising design graduate Addie Gantt working as a showroom designer for Legacy Classic. *Courtesy of Richard Bennington.*

year. She also traveled to Las Vegas (when it didn't conflict with her High Point University schedule) to help the Klaussner staff set up their showroom for the Las Vegas Home Furnishings Market.

Addie excelled in her classes and was awarded the Celia Moh Scholarship for both her junior and senior years. Addie feels that the classes that offer experiential "real-world" projects were very valuable in helping her prepare for a career in the home furnishings industry.

After her 2019 graduation from High Point University with a major in visual merchandising design, Addie secured employment in showroom design with sister companies Craftmaster Furniture and Legacy Classic Furniture that share the same High Point showroom building. In her job, she plans their market showroom designs and then helps execute them. She likes what she does and says she has found her niche. She thinks her High Point University visual merchandising design classes were extremely valuable for her employment after graduation. When I spoke with Addie, she told me, "I can't think of a thing that I learned in my High Point University classes that I haven't used on the job."

Chapter 6

SCHOLARSHIPS, ENDOWMENTS AND AWARDS FOR INDUSTRY-RELATED PROGRAMS

E ven before there was news of a furniture marketing major being planned at High Point College, local citizens who were involved in the furniture industry established scholarships to help local students get an education at the college. Currently, over thirty different scholarships have been established by furniture companies or to honor individuals with a furniture industry connection. The following are examples that show donors' enthusiasm behind industry-related scholarships. In addition to the scholarships, the Knabusch-Shoemaker Endowment provides funds for program enhancement, and the Haverty Cup recognizes an outstanding home furnishings graduate each year.

THE PATRICK H. NORTON SCHOLARSHIP: A BIRTHDAY PRESENT FOR A SENIOR EXECUTIVE

What do you do when a key executive is about to turn seventy? This was the question the La-Z-Boy Company faced in 1992, when then senior vice-president of sales and marketing Patrick H. Norton was approaching his seventieth birthday. The answer was that the entire La-Z-Boy organization, including selective members of the senior staff, sales representatives and the La-Z-Boy Proprietary Store Organization, pooled their contributions to endow the Patrick H. Norton Scholarship at High Point University for deserving students who are working toward a career in the home furnishings industry. Norton and High Point University were presented with a check for the scholarship at a banquet during the April 1992 High Point Furniture Market.

THE LAURENCE MOH AND CELIA MOH SCHOLARSHIPS

La-Z-Boy senior vice-president Patrick H. Norton. *Courtesy of Lay-Z-Boy Inc.*

Laurence Moh was a gentleman who left Shanghai, China, in 1939 with only the clothes on his back. He was fortunate enough to make it to the United States, where he studied at the Wharton School of Business at the University of Pennsylvania. Ultimately, he became very successful in the furniture business by turning his Hong Kong–based plywood company into Universal Furniture Company. Universal manufactured unassembled dining, bedroom and occasional furniture for export to the United States and other countries around the world. Mr. Moh was appreciative of his success and wanted to support outstanding young people who were preparing for a furniture industry career through scholarship assistance.

Laurence Moh, the sponsor of the Laurence Moh and Celia Moh Scholarships. *Courtesy of* Furniture Today.

For this reason, in 1999, as I was walking my two dogs down the street near my home in High Point, I received a phone call from Singapore. It was Mr. Moh, who said he was coming to High Point the following week. He wanted to make an appointment with me and anyone else I thought should be involved to discuss the possibility of establishing a scholarship at High Point University to support students who demonstrated potential to be the future leaders of the furniture industry.

We were delighted to talk with Mr.Moh, and the result of our meeting was the establishment of the Laurence Moh Scholarship, a ten-year,

Kaitlin (Thompson) Singh, a High Point University graduate and Moh Scholarship recipient. *Courtesy of Kaitlyn Singh.*

$600,000 commitment to provide a full tuition, room and board scholarship for one student per class—freshman, sophomore, junior and senior—who was majoring in an area of study that would logically lead to a career in the furniture industry. This ensured that one Laurence Moh Scholarship recipient would graduate each year for at least ten years, beginning in 2000.

As the end of the ten-year commitment was nearing, Mr. Moh ended the Laurence Moh Scholarship and created the Celia Moh Scholarship in honor of his wife, Celia. The Celia Moh Scholarship was broadened to include both High Point University and a small number of other colleges and universities with furniture industry-related curricula. Each of these colleges and universities could submit applications for the full tuition, room and board scholarship for students who were majoring in an area that would logically lead to a furniture-related career.

Kaitlin (Thompson) Singh, who graduated from High Point University in 2016, is an outstanding example of a Moh Scholarship recipient who is excelling in her career in the home furnishings industry. Kaitlin, who originally came to High Point University because she was interested in physical therapy, changed her career aspirations and graduated with a major in home furnishings marketing and a minor in marketing. She changed her major in large part because of the opportunities she saw through the "amazing High Point Market" just down the street from High Point University.

Following graduation, Kaitlin accepted employment with Klaussner Furniture Industries in her hometown of Asheboro, North Carolina. Her current position with Klaussner is in e-commerce sales. She works with e-commerce retailers such as Wayfair, Amazon, Bed Bath and Beyond and Hayneedle, as well as the e-commerce departments of traditional retailers. She is happy she chose employment with Klaussner because they are so invested in e-commerce and building an outstanding e-commerce team, which makes the company a great place for her to learn and grow. Kaitlin sees her position as being especially valuable since she is able to work with both e-commerce retailers and traditional retailers that may not be able to show everything on their sales floor. She was quick to state that Klaussner's mission is to be the leading solutions provider in the industry.

Kaitlin feels that High Point University prepared her to join the home furnishings industry. In her opinion, she got a complete education with plenty of opportunities to learn what the industry is all about from factory tours and industry professionals who spoke to the classes. The level of mentorship she felt she received as a student was second to none. Her advice to current students is for them to always challenge themselves, to expose themselves to as many facets of the industry as possible and to surround themselves with mentors they can learn from.

BILL AND BONNIE PETERSON ENDOWED SCHOLARSHIP FUND

The Bill and Bonnie Peterson Endowed Scholarship Fund was established in 2001 through contributions from Peterson's family, friends and coworkers. It was created as a memorial to Bill Peterson, the founding editor of *Furniture Today*, the leading furniture industry trade publication, in recognition of his contributions to journalism in the home furnishings industry. He was widely regarded for his integrity and understanding of the furniture industry. The scholarship, which was originally named the Bill Peterson Endowed Scholarship, was later changed to the Bill and Bonnie Peterson Endowed Scholarship to also recognize his wife, Bonnie, who was a marketing communications consultant to furniture manufacturers and a leader of the Carolinas chapter of the International Furnishings and Design Association.

GEORGE AND SHIRLEY ERATH ENDOWED SCHOLARSHIP

The George and Shirley Erath Endowed Scholarship was established through a challenge grant from longtime industry veneer supplier and board of trustees member and chair George Erath and his wife, Shirley. It is for outstanding juniors and seniors majoring in home furnishings-related areas with first preference going to children of individuals employed in the industry.

J. CLYDE HOOKER ENDOWED SCHOLARSHIP: RECOGNITION OF FIFTY YEARS WITH HOOKER FURNITURE

The J. Clyde Hooker Endowed Scholarship was established in 1996 through contributions from sales representatives, employees and friends in recognition of the contribution of Hooker's fifty years with the company. At that time, Hooker was chairman and CEO of the company. The scholarship is awarded to outstanding junior and senior students who are majoring in a home furnishings–related area, are versatile and have a high potential for service to the furniture industry.

BROYHILL FAMILY FOUNDATION ENDOWED SCHOLARSHIP

The Broyhill Family Foundation Endowed Scholarship was established in 2016 to instill designers with an understanding of basic business principles. The Broyhill Foundation of Lenoir, North Carolina, has a long record of support for High Point University. This scholarship is awarded to deserving students who are pursuing a major in the School of Art and Design and a minor in the Earl N. Phillips School of Business. The scholarship is renewable based on successful completion of the previous year's work and may be awarded to more than one student.

KNABUSCH-SHOEMAKER INTERNATIONAL SCHOOL OF HOME FURNISHINGS ENDOWMENT

Edward M. Knabusch and Edwin J. Shoemaker, the cofounders of La-Z-Boy. *Courtesy of La-Z-Boy Inc.*

The Knabusch-Shoemaker International School of Home Furnishings was established in 2002 through gifts from Patrick H. Norton, chairman of La-Z-Boy Inc.; friends and family of both Edward M. Knabusch and Edwin J. Shoemaker (who cofounded La-Z-Boy in 1927); and employees and friends of the La-Z-Boy organization. The naming of the school was done to recognize the lives and achievements of these two furniture giants.

Income from the endowment is to be used to:

- Help keep the emphasis of the home furnishings programs on student-oriented endeavors, including marketing and enrollment.
- Attract and retain top-flight faculty, distinguished lecturers and guest speakers to teach in furniture and interior design programs.
- Develop marketing initiatives that will promote the Knabusch-Shoemaker School and all programs and academic majors related to it.
- Create furniture-related summer courses, leadership seminars and continuing education offerings, which can be made available to traditional students and to persons employed in the furniture industry.
- Develop directives that strengthen the curriculum and lead to higher levels of professional accreditation and/or certification.

THE HAVERTY CUP FOR EXCELLENCE

The Haverty Cup for excellence was established by Rawson Haverty Sr. in 1988 to encourage students to pursue a career in the home furnishings industry. Haverty was a former president, chairman and CEO of Haverty Furniture, a full-service furniture retailer chain in the South and Midwest, and a past member of the High Point University Board of Trustees.

The Haverty Cup is presented annually to an outstanding graduating senior in a home furnishings–related major at High Point University. The recipient receives a cash award and a personal replica of the Haverty Cup. Their name is also engraved on the base of the Haverty Cup, which is on permanent display in the library of Norton Hall on the High Point University Campus.

Ashley (Holleran) Pratson, who received the Haverty Cup in 2010, is an example of a past recipient who is making their mark on their company and in the industry. Ashley, whose hometown is Peabody, Massachusetts, was recruited to come to High Point University to play soccer. After coming to High Point, she worked for Sherrill Furniture in their showroom at the High Point furniture markets. It was the experience of working markets for Sherrill that introduced Ashley to the furniture industry. Ashley feels that working markets for Sherrill was a unique advantage to her because it was through this experience that she discovered she wanted a career in the furniture industry.

After graduation, Ashley worked for a brief period with Haverty Furniture in one of their North Carolina stores. Deciding that retail sales was not the right fit for her, Ashley returned to High Point and took a position as a sales and marketing assistant with Miles Talbott, a custom upholstered furniture manufacturer located in the city. While working for Miles Talbott, she took evening graduate business classes, completing her master of business administration (MBA) degree at High Point University in 2013. Ashley subsequently married a man who she met at High Point University who is also from New England. Eventually, they decided they wanted to move back to New England to be closer to their families. Being progressive and open to new ideas, the president of Miles Talbott agreed to let Ashley continue working with the company while actually living in Boston. Ashley, whose title is now vice-president of merchandising and strategic marketing for Miles Talbott, works remotely from her home in downtown Boston through the use of her computer and conference calls. She spends a lot of time on the phone with colleagues, especially before market. This arrangement is working well, despite it forcing her and her colleagues to be more organized and have a strict schedule so that they can meet their deadlines.

Ashley flies to High Point almost every month and also travels extensively throughout the country. She works with the Miles Talbott sales force by

Left: The Haverty Cup, which is awarded annually to an outstanding senior with a home furnishings–related major. *Courtesy of Richard Bennington.*

Right: Ashley (Holleran) Pratson, a Hight Point University graduate and 2010 Haverty Cup recipient. *Courtesy of Richard Bennington.*

representing the company in retail sales training and working with the dealers on floor samples and properly organizing their retail sales floors. She feels that traveling to different sections of the country has given her a unique insight into the widely varying needs of the marketplace. She also works with the company's creative director by shopping for fabrics at the High Point Showtime Fabric Fair and helping design the company's High Point furniture market showroom.

Ashley has found something she loves—both the creative and business aspects of her job. And she can be with her family in Boston while still keeping her ties with her furniture family in High Point. She feels that the many opportunities she had for networking, beginning at the intern level, were unique for High Point University students. She also thinks that it is very important for current HPU students to take business classes and to get a master's degree, if possible.

Chapter 7

NORTON HALL

An Industry-Sponsored Building Appears on Campus

By 1999, there had been 335 graduates of the Furniture/Home Furnishings Marketing Program. Approximately 75 percent of these graduates were working in jobs related to the furniture industry with suppliers, manufacturers, retailers or interior design firms. At that time, the furnishings-related classes were largely held in a classroom on the first floor of Cooke Hall (now Norcross Hall), and the interior design classes were held in a classroom/studio on the main floor of the old student center. With the success of graduates finding employment and the clear need for more facilities, the Home Furnishings Advisory Board, chaired by Clarence Smith of Haverty Furniture, unanimously passed a motion by board member Joe Carroll, publisher of *Furniture Today*, to raise funds for a building on the High Point University campus to house the furniture industry–related programs.

After a series of additional meetings, it was determined that there was sufficient interest to continue with this project, and Mercer Architecture, a High Point–based architecture firm, was selected to come up with a design for the building. The result was a twenty-seven-thousand-square-foot, three-story, Georgian-style building designed to fit in well with the overall architecture of the campus. The location was to be just to the right of the Hayworth Chapel on the High Point University campus. At this stage, the name of the building was the "International School of Home Furnishings at High Point University."

Norton Hall. *Courtesy of High Point University.*

LET'S GET GOING—IT IS TIME TO RAISE THE MONEY

A fundraising program was organized under the direction of John Lefler, High Point University vice president for institutional advancement, to raise $2.6 million, the projected price tag for the building. Every area of the three levels of the building and the building itself were naming opportunities. Potential donors could have their name placed on any area of the building, including entire floors, atriums, classrooms, computer labs, a tiered lecture hall, the building's library, offices or even the elevator. I was impressed with the fundraising plan Lefler proposed to contact the "movers and shakers" throughout the home furnishings industry. Based on information from the advisory board and other industry leaders, Lefler laid out an elaborate plan for determining who the best donor prospects were and who would be the best person to contact for each of them.

Among the talking points for the presentations were:

- The industry was becoming more sophisticated. Competition from other industries was such that household furnishings garnered only about 1 percent of consumer income, and there was a potential for this percentage to be considerably greater.
- The amount of imported home furnishings was running about four times more than the amount of merchandise manufactured in the United States.

- Changes in distribution were occurring every day. According to *Furniture Today* publisher Joe Carroll, there were, at that time, over sixty-four different types of outlets selling home furnishings, which could mean more job opportunities for HPU grads.
- The secret of success in the future would be to build organizations staffed with knowledgeable individuals who knew the industry yet were flexible enough to adapt to change.
- One of the best ways to have access to these young individuals was to support the International School of Home Furnishings at High Point University. HPU had built and was constantly refining a practical curriculum to educate the industry leaders of tomorrow. When this fundraising campaign was completed, High Point University would have the physical facilities to properly support this educational program.

HPU Names the Building "Norton Hall"

One of the industry leaders we contacted early in our fundraising was Patrick H. Norton, chairman of the board of La-Z-Boy Inc. Norton had always been a friend to High Point University but had been even more so since the establishment of the Patrick H. Norton Scholarship at HPU as a gift for his seventieth birthday. But why was Norton so persistent in his generosity to the university? A quick look at Norton's background provides some insight into some of the possible reasons for his interest in High Point University.

In the summer of 1940, when Norton was barely eighteen years old, he convinced his mother to sign for him to join the army air corps, and there, he served with distinction in World War II. Following his military service, Norton began working in the furniture industry. Beginning in retail, he worked from the ground up, eventually becoming vice president of sales and marketing for Ethan Allen. It was during his time at Ethan Allen that the company pioneered the in-store gallery concept of selling home furnishings products in Ethan Allen retail stores. This allowed consumers to buy completely coordinated room packages, including all of the furnishings and accessories. Consumers were also given specifications for the paint colors to be placed on the walls.

The success of this Ethan Allen program caused the management of La-Z-Boy to notice Norton's marketing insight and invite him to join their organization. Because of his military service and because he had directly entered the world of business after leaving the armed forces, Norton did not attend a college or university that he could call his alma mater. What Norton saw going on at High Point University and its relationship with the furniture industry is certainly one of the main reasons his interest in the university began to increase. In some ways, I think High Point University became his substitute alma mater.

By the time of our request, Norton had been named to the High Point University's board of trustees. He responded favorably to our request for a financial commitment to the building, and because of that, the board of trustees voted unanimously to name the building Norton Hall in his honor. High Point University had also recently awarded Norton an honorary doctorate degree.

THE INDUSTRY JUMPS ON BOARD

Within a relatively short period, through the efforts of members of the furniture advisory board and other individuals with influence in the furniture industry, the fundraising campaign was successful. When the funds raised were at $2.1 million (out of the $2.6 million projected total cost for the building) and the furniture advisory board and others were contemplating on how to close the gap, Joe Carroll volunteered to ask Laurence Moh for a commitment, thinking he might contribute $50,000 toward the building. When Carroll explained the situation to Mr. Moh and asked him if he would be willing to contribute toward the building, Moh asked, "How much money do you need to complete the project?" Carroll told him that $500,000 was needed. Moh simply said "OK," and wired the funds to High Point within the next couple of days. The university subsequently decided to name the mid-level of the building for Moh. Not wanting credit for himself, Moh asked that the mid-level of the building be named in memory of Bill Peterson, the founding editor of *Furniture Today*.

With Moh's contribution, the fundraising campaign was completed. Every naming opportunity in Norton Hall was taken by a donor, and there was a name for each floor, classroom, office and most of the other areas in the building. In addition, most of the donors provided funds to furnish their area of the building. It is gratifying that almost all segments of the industry, from manufacturers, to retailers, wholesalers, suppliers and other industry professionals, are represented in the plaques on the walls of the building. The Kohler Company, owner of Baker Furniture Company, contributed the fixtures for the restrooms on each of the building's three floors.

DONORS BY AREA OF THE FURNITURE INDUSTRY

Manufacturers: La-Z-Boy Inc., Broyhill Furniture, Century Furniture, Sealy Inc., Bassett Furniture, Furniture Brands International, Sherrill Furniture, Thayer Coggin, Industrie Natuzzi, Klaussner Furniture, Vaughan Furniture, Ladd Inc., Stickley, Richardson Brothers and the Rowe Companies.

Retailer: Haverty Furniture.

Wholesaler: Huntington Wholesale Furniture.

Suppliers: Barnhardt Mfg. Co., Finch Industries., Culp Inc., Leggett and Platt, Erath Veneer Co.

Individuals: Bill Peterson (furniture trade publication), Laurence Moh (manufacturer/importer), Robert Gruenberg (furniture market), George Erath (furniture supplier), Gene Kester (sales representative), J. Smith Young (manufacturing), Bob Timberlake (artist/licensee), Mr. and Mrs. Sherrill Shaw (retailing), and Mr. and Mrs. Charles Greene (manufacturing) and Bill and Jean Kester (retailing).

Table 1 (*opposite*) shows the named areas of the building by floor and donor.

A brief description of three of the specific rooms provide a hint of the variety and practicality of the different areas within Norton Hall.

NORTON HALL LECTURE ROOM

The Norton Hall tiered lecture room was designed to seat sixty-six people and has the latest in audiovisual capabilities, so it can be used as a classroom and as a meeting place for a variety of groups both from the university and the home furnishings industry.

Table 1

Named areas of Norton Hall

The following are named areas of the building by floor and donor:

Lower level

Floor in memory of Krell B. Norton	Mr. and Mrs. Pat Norton
Tiered Lecture Hall	La-Z-Boy Foundation
Multi-purpose Lecture Hall	Broyhill Family Foundation
Classrooms	Haverty Furniture Co. Inc. Century Furniture Co.
Student Lounge/Faculty Offices	Sealy, Inc.

Mid-level

Floor in memory of Bill Peterson	Laurence and Celia Moh Foundation
Library	Bassett Furniture Industries, Inc.
Classroom in memory of Bob Gruenberg	George S. Erath Family
Classroom	Gene Kester
Atrium in Memory of J. Smith Young	Bob Timberlake & Assoc
Vestibule	Barnhardt Mfg. Co
Offices	Huntington Wholesale Furn. Co. Finch Industries Mr. & Mrs. Sherrill Shaw Mr. & Mrs. Charles Greene Industrie Natuzzi SpA Klaussner Furniture Industries, Inc. Vaughn Furniture Co., Inc. Bill & Jean Kester Ladd, Inc.
Conference Room in memory of Alfred Audi	Stickley Furniture, Inc

Upper level

Floor	Culp, Inc.
CADD Lab	Richardson Industries
Interior Design Studio	Leggett & Platt, Inc.
Interior Design Studio	Sherrill Furniture Co.

Table 1: Named areas of Norton Hall by floor and donor. *Courtesy of High Point University.*

Norton Lecture Hall, sponsored by the La-Z-Boy Foundation. *Courtesy of Benita VanWinkle.*

NORTON HALL COMPUTER LAB

This computer lab is one of three classrooms containing computers in the building, which reflects the faculty's commitment to focus on CAD and other computer software training in a variety of situations so that graduates will understand the many uses of computers in actual businesses. This will help the students to be prepared for how the world of work will be in the future.

BASSETT FURNITURE LIBRARY

The Bassett Library in Norton Hall houses a carefully selected collection of furniture- and design-related books and other materials. In addition, there is student space for studying, and the library accommodates small meetings. An extra bonus is that the library also houses Pat Norton's collection of awards and other professional materials, which have proven to be good historical information on the home furnishings industry.

The students in the image at the top of the following page, with their designer hard hats, demonstrate their appreciation of the realities of being safe on construction sites. But at the same time, they are able to show their design talents which, hopefully, reinforces their commitment to becoming design professionals.

Interior design students, shown in their designer hardhats, looking up from a lower level of Norton Hall. *Courtesy of Cathy Nowicki.*

Bassett Furniture Library in Norton Hall. *Courtesy of Richard Bennington.*

Chapter 8

MORE FURNITURE INDUSTRY–RELATED NAMES SHOW UP ALL OVER CAMPUS

A number of prominent furniture companies and families have stepped up to enhance the High Point University campus experience through financial commitments, which have resulted in their names being placed on buildings and endowments. These are also tangible pieces of evidence of a strong partnership between the university and the furniture industry.

THE HAYWORTH FAMILY STEPS UP AGAIN

The Charles E. and Pauline Lewis Hayworth Fine Arts Center

The Charles E. Hayworth Sr. Chapel was followed several years later by another prominent building on the High Point University campus that was also named for Hayworth family members. This building, the Charles E. and Pauline Lewis Hayworth Fine Arts Center, named for Charles Jr. and his wife, Pauline, opened in 2002. The Hayworth Fine Arts Center contains a five-hundred-seat performance hall, music lab, art gallery, dressing rooms, costume shop, art studio, darkroom and faculty offices.

Charles and Pauline Lewis Hayworth Fine Arts Center. *Courtesy of High Point University.*

David R. Hayworth Hall, College of Arts and Sciences and Park

Charles's brother David made the Hayworth name even more noticeable on campus when he made the lead gift for the David R. Hayworth Hall, a classroom and office building for the Departments of Religion, Philosophy and History. This building, which opened in 1998, is conveniently located next to the Hayworth Memorial Chapel.

David Hayworth subsequently made two more gifts, resulting in naming opportunities for the Hayworth family. The High Point University College of Arts and Sciences was named the David R. Hayworth College of Arts and Sciences in his honor. The David R. Hayworth College of Arts and Sciences includes a relatively large number of departments, such as English, French, Spanish, religion, philosophy, history, political science, human relations, psychology, sociology and anthropology, music, mathematics and theater.

The third area on campus named for David Hayworth is the David R. Hayworth Park, located prominently on the campus between the R.G. Wanek Center and the Slane Student Center. It contains fifteen-foot-

David R. Hayworth Hall. *Courtesy of Richard Bennington.*

David R. Hayworth Park. *Courtesy of High Point University.*

Finch Residence Hall. *Courtesy of High Point University.*

tall waterfalls that students can walk behind, a variety of sculptures and botanical gardens. This park is used for outdoor concerts, student meetings and, occasionally, even classes.

FINCH RESIDENCE HALL

This forty-three-thousand-square-foot men's residence hall was dedicated in 1990 to honor the grandchildren of Meredith Slane (Finch) Person and her late husband, Tom A. Finch. Mr. Finch was president and chairman of the board of Thomasville Furniture Industries and later the senior vice-president of Armstrong World Industries. The four-story residence hall contains 111 double rooms, with lounges on each floor and a large first-floor lounge with an adjacent serving kitchen.

GENE AND JANE KESTER INTERNATIONAL PROMENADE

In 2006, the international promenade of High Point University was named in honor of Gene Kester and his wife, Jane. Prior to his retirement in 2008, Gene Kester spent his career in the furniture industry, first as part owner of Rose Furniture, a longtime High Point retailer, followed by a successful career as a manufacturer sales representative. He has been an active supporter of High Point University, including serving as chair of the university's board of visitors and board of trustees.

The Gene and Jane Kester International Promenade is lined with international flags, which represent the dozens of nationalities that make up High Point University's student body. An "outdoor classroom," the promenade also features wireless internet service, classical music that plays throughout the day and twenty sculptures of famous historical figures that can be found along many of the promenade's benches. Many Instagram posts feature an HPU student or visitor posing next to a life-sized bronze likeness of historical heroes ranging in time from Socrates to Rosa Parks.

Gene and Jane Kester International Promenade. *Courtesy of High Point University.*

Norcross Graduate School. *Courtesy of High Point University.*

NORCROSS GRADUATE SCHOOL

The Norcross Graduate School building has multiple uses, as it houses the offices for the graduate school faculty and staff, the Information Technology Center, computer science laboratories, classrooms and offices for the Departments of English, Criminal Justice, Human Relations, Sociology, Nonprofit Studies and Exercise Science. The Norcross Graduate School offers a number of graduate degrees, including a master of arts in strategic communications, a master of business administration, as well as several masters- and doctoral-level programs in education and health sciences.

The building was dedicated in 2006 and named in honor of Mark Norcross. Mark Norcross is the president, founder and chief executive officer of Mark David Inc., a High Point–based manufacturer and marketer of high-end hospitality furniture for luxury hotels, such as Ritz Carlton, Four Seasons and Wyndham. When Mark David, Inc. was acquired by Kohler Interiors in 2008, Norcross's duties increased to include other Kohler divisions, such as Ann Sacks Tile and Stone, Inc. and Baker Kallista Plumbing, helping the Kohler Company become the leader in the hospitality furnishings market worldwide.

PLATO S. WILSON SCHOOL OF COMMERCE

The construction of the Plato S. Wilson School of Commerce building was made possible through a generous gift from longtime furniture manufacturer representative Plato Wilson. Wilson, prior to his retirement in 1990, was an extremely successful furniture salesperson, working for only two companies, Henredon Furniture Company and, later, Henkel-Harris Furniture Company. He partnered with his retail store customers and trained their salespeople so well that he overwhelmed the competition. During his career, he traveled 2.6 million miles and had life sales of $154 million, including several $10 million dollar sales in one year, and he even scored $1 million dollars in a day! In recognition of his achievements in the furniture industry, he was inducted into the American Furniture Hall of Fame in 2004.

The sixty-thousand-square-foot building was completed in 2009 and houses classes in accounting, finance, marketing and management. It contains specially designed physical spaces for business endeavors, such as a virtual stock trading room, a large boardroom and a convention-type ballroom with full kitchen facilities.

R.G. WANEK CENTER, WANEK SCHOOL OF NATURAL SCIENCES AND *MENTORING 2000* SCULPTURE

R.G. Wanek Center

The 277,000-square-foot University Center, built in 2009, was later renamed the R.G. Wanek Center in recognition of a gift from the Ronald and Joyce Wanek Foundation of Arcadia, Wisconsin. Ronald Wanek started manufacturing furniture with only thirty-five employees in 1970. His company, Ashley Furniture Industries Inc., with corporate headquarters in Wisconsin, is now the largest furniture retailer in the United States and is one of the world's largest furniture manufacturing and distribution facilities, employing twenty-seven thousand people worldwide. Ashley Furniture has a major commitment in North Carolina through its presence at the High Point Market and its distribution center in Advance, North Carolina, just a short drive from High Point.

Speaking on behalf of the Wanek Foundation, Ron Wanek praised High Point University for its educational efforts in entrepreneurship, technology,

Plato S. Wilson School of Commerce. *Courtesy of High Point University.*

R.G. Wanek Center. *Courtesy of High Point University.*

business and the free enterprise system. To quote Wanek from a statement released by High Point University that appeared in the *Greensboro News and Record* and other local newspapers in August 2013: "The educational emphasis on communication and leadership that is the focus of High Point University is providing essential skills to tomorrow's leaders." These were cited as some of the reasons the Wanek Foundation supported High Point University. The Wanek Center includes residential space for 580 students, a twenty-four-hour study space and satellite library and several dining options, including a restaurant on the fourth floor that serves as a learning lab for students practicing etiquette and job interview skills. It also includes a fully functioning movie theater, a convenience store and a gaming center/recreation area.

Wanek School of Natural Sciences

The Wanek name became even more prominent when High Point University's board of trustees, in January 2018, voted to name the new School of Natural Sciences in honor of Todd Wanek and his wife, Karen. Todd is the CEO of Ashley Furniture and the son of Ron Wanek, founder of Ashley Furniture and benefactor to the Ronald and Joyce Wanek Foundation for which the university's Wanek Center is named. Karen

Wanek School of Natural Sciences. *Courtesy of High Point University.*

Wanek is president of Superior Fresh, which raises Atlantic salmon and rainbow trout through sustainable agriculture.

The 128,000-square-foot Wanek School of Natural Sciences houses the majors of biology, chemistry, biochemistry, neuroscience and physics. The building is dedicated to experiential learning and features four stories of innovative lab and classroom space, as well as the Culp Planetarium.

Culp Planetarium

The partnership of Culp Inc. and the Culp family with High Point University became prominent when Robert Culp III (Rob) and his wife, Susan, contributed substantially to the new planetarium that has recently been built on the High Point University campus. The planetarium, located on the first floor of the Wanek School of Natural Sciences building, has been named the Culp Planetarium in their honor. This three-story planetarium and lecture room space includes 125 seats, a fifty-foot-tall dome and equipment to support the teaching of courses such as astronomy, earth science and anatomy, as well as supporting the biology, chemistry and physics majors. The university also serves High Point families and area schools by opening the planetarium periodically for free public shows.

The Culp Planetarium. *Courtesy of High Point University.*

Wanek Mentoring 2000 sculpture. *Courtesy of Rebecca Slife.*

Mentoring 2000 *Sculpture*

The *Mentoring 2000* sculpture, which was designed by Ron Wanek and donated to HPU by the Wanek family, depicts Wanek and his son, Todd, mentoring three of his grandchildren in the furniture industry by building a piece of furniture. The university chose to place the sculpture in front of HPU's Stout School of Education because the focus of the education school is to help create mentors.

WEBB CONFERENCE CENTER AND WEBB SCHOOL OF ENGINEERING

Mark Webb, a 1983 High Point College graduate, honored his alma mater with two major gifts, which have resulted in the Webb family name being placed on the High Point University campus conference center and the new school of engineering. Mark owns and operates Interstate Foam and Supply Inc. in Conover, North Carolina, a company that was founded by Mark's father, Lewis, in 1981. Under Mark's leadership, the company

Mark and Jerri Webb, benefactors, in front of Webb Conference Center. *Courtesy of High Point University.*

has grown to be a major fabricator and distributor of quality seating and polyurethane foam components for the home furnishings industry. Interstate Foam and Supply remains in Conover on a fifty-two-acre site, with three manufacturing plants employing more than four hundred associates. The gifts were made in honor of Mark's late father, Lewis, his mother, Janice, and the entire Webb family.

The Webb Conference Center is a multipurpose facility conveniently located on International Avenue, near the major downtown entrance to the campus. The Webb Conference Center is used by students, faculty and other members of the High Point University community for classroom lectures, meetings, banquets and other events.

The Webb School of Engineering, located on the High Point University campus in the recently renovated Couch Hall, currently offers degrees in computer science, with concentrations in cybersecurity, software and systems and visual computing; additional majors are planned for the future. In the news release announcing the establishment of the Webb School of Engineering, Mark Webb said, "As HPU prepares for the next generation of leaders, my family and I are honored to support the future of HPU students and the values of God, family and country HPU represents." Because of his generous support and donations to High Point University, Mark Webb was named Alumnus of the Year for 2018. HPU president Nido Qubein often describes the Webb family as a symbol of the American dream.

HARRIS SALES EDUCATION CENTER

Housed in Cottrell Hall on the High Point University campus, the Harris Sales Education Center features a large lobby, three specialized sales and

Jeff and Jason Harris, benefactors of the Harris Sales Education Center, in Cottrell Hall. *Courtesy of High Point University.*

interview rooms and the Free Enterprise Conference Room. Inside the labs, which mimic three real-world work settings, including a tech company, a financial firm and a healthcare setting, students can record themselves selling to clients or practicing mock job interviews while also recording the reaction of the person to whom they are selling.

The Harris Sales Education Center is named in honor of Darryl and Stella Harris, who, in 1969, founded Furnitureland South, which has grown to be the world's largest retail home furnishings store. The dedication plaque honoring the Harrises stresses "their love of God, dedication, hard work, capacity to dream and to realize great things, dedication to family, friends, associates, and the high regard for the sales profession." Their two sons, Jeff and Jason, are now co-owners and share in the management of the company.

When Jeff, a 1990 graduate of High Point College/University and furniture marketing major, was asked about why he and his brother chose to make the financial commitment for the Harris Sales Education Center, he said they wanted to extend the legacy of excitement, energy and passion that they sense is present at High Point University. He stated that his own major in furniture marketing allowed him to extend his knowledge from furniture retailing, which he had learned while growing up, to an in-depth knowledge of how the other parts of the industry work. It was obvious that he values what he learned from factory tours and other industry components, such as manufacturing and design.

He stressed that the life skills students learn at High Point University are extremely valuable. An example he gave is the university's emphasis on teaching students how to present themselves in a variety of settings. He also stressed that today's graduates should be exposed to business analytics. In other words, he thinks that it is important for students to learn to gather data, assimilate the data and use it in decision-making.

The commitments of these individuals from the furniture industry, both alumni and non-alumni, reflect the philosophy that High Point University is a forward-thinking, progressive institution dedicated to providing the type of education that college graduates will need to succeed and become leaders in their chosen career fields.

Chapter 9

THE TIMES THEY ARE "A-CHANGING"

*More Imports, Less U.S. Manufacturing, Online Shopping
and the High Point Market Becomes the
Largest in the World*

Modern furniture manufacturing, in many ways, began with the end of World War II and the many advances that were made in the 1950s. Chapter 2 covered a number of reasons for the timing of these advances, the greatest of which was the housing shortage after World War II. This shortage resulted in a postwar housing boom that, of course, meant a large number of new houses were constructed and purchased using funds from the GI Bill. As a result, times were good for furniture manufacturers and companies in and around High Point.

SAILING ALONG—THEN THE WIND CHANGES DIRECTION

The prosperous times for the furniture manufacturers in and around High Point lasted well into the 1970s. But a saying that I have heard quoted is, "In life, we can be sailing along, but then the wind just changes direction," and this seems to apply to furniture manufacturing in the area around High Point, as well as other parts of the United States. Now let's look at the changes that have affected the furniture industry.

A container ship bringing imported products from the far east to the United States. *Courtesy of Shutterstock.*

OH, MY—HERE COME THE IMPORTS

The main factor that caused "the wind to change direction" for the U.S. furniture manufacturers was furniture that was formerly being produced domestically was being imported into the United States from outside of the country. But once the imports started, the change in wind direction was relentless. Jerry Epperson, one of the founders and managing directors of Mann, Armistead & Epperson Ltd. in Richmond, Virginia, has discussed in considerable detail, through industry presentations and the *Furnishings Digest* newsletter, how imports have increased since the 1970s. Specifically, imports began arriving in seemingly ever-increasing quantities from China, Vietnam and other offshore destinations.

These imports, in some ways, can be looked at as part of a continuing cycle. In the early to mid-1900s, furniture manufacturing shifted from areas like Grand Rapids and Chicago to North Carolina and other areas of the South in search of lower labor and other production costs. Then furniture manufacturing was shifted offshore, largely for the very same reasons.

In 1979, imports made up less than 9.1 percent of the furniture products available in the United States, and the largest source nations were Taiwan, Canada, Denmark, Italy and Yugoslavia. In the 1980s, 1990s and 2000s, imports continued to arrive in increasing quantities in almost every product category, including occasional tables, dining tables and chairs, bedroom furniture, leather and fabric upholstery and metal dining room furniture. Although furniture had begun arriving from many different countries around the globe by the 2000s, the three largest import sources were China, Vietnam and Canada. To illustrate the volume of imports into the United States, 75.5 percent of all the household furniture that was sold in 2017 was imported from outside the country. To break it down into product categories, table 2 shows the percentage of wood furniture, upholstered furniture and metal furniture sold in the United States in 2017 that was imported. The table also shows the dramatic increase in the amount of imports by comparing these same categories of products in the United States in 1993 and 2017. These numbers tell the story of increasing quantities of imported furniture being sold in the United States.

Table 2

Import Share of Household Furniture Sold at Retail in the US

PRODUCT TYPE	1993	2017
Wood	24.8%	88.9%
Upholstery	5.7%	49.1%
Metal & Other	23.3%	85.3%

Source: Mann, Armistead & Epperson

Table 2: Import share of household furniture sold at retail in the United States. *Courtesy of Mann, Armistead & Epperson.*

FACTORIES CLOSE AND COMPANIES CHANGE FROM MANUFACTURERS TO MARKETERS

Regardless of the increased complexity resulting from dealing with offshore sources having different shipping requirements, languages and cultures, many furniture factories in North Carolina and other areas of the South have closed because of the cost differential in retailers selling imported products as opposed to domestically produced products. Imports are simply cheaper.

Although it may not have been evident in the beginning, many American furniture companies found another reason to close their factories and buy from offshore sources: it provides greater flexibility as to the variety of furniture they are able to provide to U.S. consumers. As long as they were producing in their own factories, they were constrained by aging facilities and the types of equipment available for use in those factories.

Now, these companies have more freedom to provide whatever products they perceive will be popular with consumers. In other words, they have become marketers of products manufactured in factories located in China, Vietnam, Canada, Mexico and elsewhere around the globe. If their market research seems to indicate that a particular style will be popular with consumers, they can probably find a company to make it somewhere outside the United States, often at a lower cost.

CHANGES IN THE WAY AMERICA SHOPS

I remember calling the local Sears store in the 1960s and 1970s, and the person answering the phone would say, "Sears, where America shops. How can I help you?" Well, America did shop at Sears, either from the Sears catalog or in the Sears store for a long period of time. To underscore my point, the father of one of my college roommates in the 1960s was the manager of the Sears store in Fredericksburg, Virginia, which, at the time, I thought was a really good job. But just like the furniture industry, Sears was "sailing along, and the wind changed direction," resulting in the American public deciding to shop elsewhere.

Let's look at the changes in how the American public has shopped for furniture. The furniture industry has always been "fragmented and diverse," meaning that there are many different types of products manufactured by a wide array of companies, and they are offered for sale through many different outlets at various price ranges and quality levels. Even though shopping patterns vary from one geographical area to another, there are definite trends as to where and how the American public shops for furniture.

Sears retail store of the 1950s to 1970s. *Courtesy of* Furniture Today.

To go back to post–World War II, independent, single-market retailers were popular places to shop, as were the "mass merchandisers"—Sears, JCPenney and Montgomery Ward. These "mass merchandisers" began as catalog retailers. Shoppers could look through the Sears, JCPenney or Montgomery Ward's catalogs at their leisure and decide what they would like to purchase. Eventually, these companies built retail stores, where consumers could see, touch and feel the products that they may have first seen in the catalogs before making their final decision. In larger metropolitan areas, department stores were popular places for consumers to shop, especially among those who were shopping for the latest in-home fashions. Stores with names like Macy's, Bloomingdale's and Marshall Field were definite go-to places for many consumers.

In the late 1960s and 1970s, very large warehouse showrooms with names like Levitz, Wicke's, Mangurian and Gold Key became popular. These were stores with large showrooms displaying a wide selection of home furnishings, and usually in the same building, there was a warehouse where the "back up" inventory for the merchandise displayed on the sales floor was stored. These stores seemed to be saying, "We have everything you could possibly want—and it is in stock. Bring your pick-up truck, make a decision and take it home that very same day." These stores that specialized in home furnishings were followed up by general merchandise showrooms, like Sam's and Costco, which offered a selection of home furnishings as part of their product mix.

Throughout the 1980s and 1990s, the number of different outlets where consumers could buy furniture was increasing. Lifestyle stores like Pottery Barn and Ikea, a Scandinavian lifestyle store featuring low-priced, ready-to-assemble furniture, seemed to appeal to the younger customers who wanted a "now" look in their homes.

Beginning with Ethan Allen in the 1950s, manufacturer-branded stores began opening in the United States. Today, in addition to Ethan Allen, there are a number of other manufacturer-branded stores, like La-Z-Boy, Bassett and Ashley, in many parts of the country. These are either owned by the manufacturer or are operated by another company as a franchised store.

Perhaps the most dramatic current trend in the retail home furnishings marketplace began unfolding as younger consumers who grew up with the internet, computers and cell phones began buying home furnishings. These consumers, mostly born after 1980, are so familiar with these "technological wonders" that buying tables, chairs, bedroom furniture and even mattresses from online websites is almost second nature to them. Current writers in

Ashley Furniture HomeStore: an example of a manufacturer-branded franchised retail store. *Courtesy of Richard Bennington.*

A consumer shopping for furniture online. *Courtesy of Benita VanWinkle.*

the furniture industry press talk about some of these consumers practicing "showrooming." Showrooming is shopping for furniture in local retail stores and then seeing if they can get that particular piece of furniture—or something very much like it—cheaper from an internet retailer. The definition of "where America shops" is definitely changing.

THE HIGH POINT FURNITURE MARKET

Now, let's switch our focus from discussing the furniture industry in general to talking specifically about the furniture market. As I referred to briefly in chapter 4 when discussing my first visit to market, the High Point Furniture Market first came to national prominence as the "Southern Exposition" or, as most people referred to it, the "Southern Furniture Market." It was 150 miles long, stretching from factory showrooms in Hickory and Lenoir on one end to showrooms in High Point and Burlington on the other end. The two largest showroom buildings of the market were the Hickory Furniture Mart in Hickory and the Southern Furniture Exposition Building in High Point. Although the High Point end of the market was much larger, I have heard it said that because there was a shortage of hotel rooms, some buyers would shop in Hickory and Lenoir while others shopped in High Point. Then they would switch with those in High Point who were going to Hickory and vice versa, making more certain there would be enough hotel rooms for all the market visitors as they completed their market shopping.

Other changes were also occurring in the Southern Furniture Exposition. Throughout the 1950s, the main markets were held in January and July of each year; however, informal "in-between" or "mid-season" markets began to emerge in April and October. By 1960, the attendance of the April and October markets had begun to surpass the January and July markets. This trend continued until 1982, when the January and July markets were discontinued. And by 1985, the Southern Furniture Market was consolidated in High Point, as the major manufacturers that had been showing their furniture in showrooms in Hickory and Lenoir moved their showrooms to High Point.

WOW, THE MARKET IS CHANGING FROM REGIONAL, TO NATIONAL, TO INTERNATIONAL

The High Point Furniture Market had, within a relatively short period of time, moved from being a regional market, showcasing primarily home furnishings products manufactured in North Carolina and adjoining states, to being the largest furniture market in the United States and, more recently, the largest home furnishings market in the world. In 1989, to reflect this change, the Southern Furniture Market was renamed the International Home Furnishings Market, and the name of the Southern Furniture Exposition Building was changed to the International Home Furnishings Center (IHFC).

The international aspect of the market was revealed not only by exhibitors from countries like China, Vietnam, Canada and Mexico, but also by the fact that many of the developed countries throughout the world were represented in the thousands of buyers and other visitors who attend the market. By 2001, market organizers estimated that eighty thousand visitors, on average, were attending the April and October markets. In 2006, the International Home Furnishings Market officially changed its name to the High Point Market: the World's Home for Home Furnishings.

The International Home Furnishings Center (IHFC), the largest showroom building in High Point. *Courtesy of International Market Centers.*

HIGH POINT MARKET AUTHORITY PROVIDES NEEDED ORGANIZATION TO THE MARKET

In 2001, the High Point Market Authority was organized to improve the experience for those visiting the market. In realizing the positive financial impact the furniture market has on the state, the North Carolina General Assembly worked out a plan to provide funding for the High Point Market Authority and its market services. Some of the services provided by the High Point Market Authority are:

- **Centralized Registration**: A centralized registration system to allow visitors to obtain a single badge for all major market venues.
- **Transportation Terminal and Free Shuttles**: A transportation terminal from which free shuttles can be taken to major airports, hotels and showrooms in the area.
- **Marketing**: The High Point Market Authority is the overall voice of the market, including the market's official website, press releases, coverage, using various types of media,

High Point Furniture Transportation Terminal, organized and supervised by the High Point Market Authority. *Courtesy of High Point Market Authority.*

including social media and maintaining an on-site media center during the market.

- **Online Market Planning Tool**: This tool allows visitors to plan their market visit before getting to High Point.
- **Information Booths**: These booths are strategically located in the center of the market, where visitors can obtain information and assistance to help make their market visits easier and more productive.
- **Services for International Visitors**: This includes such services as a lounge for relaxation, foreign currency exchanges, interpreter referrals, a complimentary business center and a prayer and meditation room.

INTERNATIONAL MARKET CENTERS BUY HIGH POINT MARKET BUILDINGS

In 2011, a new company, International Market Centers LP, was formed to purchase much of the showroom and exhibition space in both High Point and Las Vegas. The result was the creation of the world's largest network of showrooms and exhibition spaces for the home furnishings, gift and home décor industries. Today, International Market Centers (IMC) owns and operates approximately 7 million square feet of showroom space in High Point, containing 70 percent of the exhibitors at the market. IMC also purchased AmericasMart Atlanta, which has resulted in the company operating a total of approximately 20 million square feet of showroom and exhibition space in High Point, Las Vegas and Atlanta.

What this has meant for High Point is an infusion of capital to modernize the major downtown showroom buildings, like the International Home Furnishings Center, to create a world-class shopping experience for buyers and a venue for exhibitors to show their products. In addition to updating the showroom buildings and making them equal to all types of wholesale markets around the world, International Market Centers has partnered with the High Point Market Authority to reach buyers and designers around the United States and beyond, thereby keeping the image and appeal of the High Point Market strong. Using a combined approach, IMC and the High Point Market Authority have been able to offer services that individual exhibitors or buildings could not offer by themselves.

International Market Centers and High Point Market Authority combine to offer educational seminars during market and to maintain contact with retail buyers, designers and other potential market visitors between markets using such techniques as blogs, Instagram and Pinterest.

Chapter 10

HIGH POINT COLLEGE UNDERGOES A COMPLETE TRANSFORMATION

From a Small Methodist College to a Larger, Dynamic Life Skills University

In chapter 1, in the discussion of the founding of High Point College, I referenced the book titled *Remembered Be Thy Blessings* by Dr. Richard McCaslin, which covers the period of HPU's history from 1924 through 1991 (the "college years"). On reading further, it is clear that High Point College encountered many of the same challenges that were faced by other four-year liberal arts colleges throughout the United States: how to stabilize enrollments, build endowments, attract faculty with the appropriate credentials, meet accreditation standards and build a physical infrastructure where students would want to live and study during their college careers. High Point College presidents, along with their respective boards of trustees, have tried to solve many of these problems in a variety of ways. The following is a snapshot of a few of the techniques used by a number of different administrations to try to strengthen the college.

USING REAL ESTATE INVESTMENTS TO HELP BUILD THE ENDOWMENT

Dr. Wendell Patton, the president of High Point College from 1959 to 1980, and his board of trustees, decided to try and build the institution's endowment through investing in local real estate. Their most significant real estate investment was a piece of property in downtown High Point called, by

Holt McPherson, chair of the board of trustees, the "Magic Block." This property, now owned by International Market Centers and occupied by the Showplace furniture market building and parking lot, was, for several years, the site of the High Point Sears store. This apparently turned out to be a good investment because the college received a percentage of the store's net sales as rent.

Dr. Wendell Patton, president of High Point College from 1959 to 1980. *Courtesy of High Point University.*

Another real estate project during Dr. Patton's presidency was building a shopping center at the intersection of Lexington and Montlieu Avenues, several blocks east of the campus in what is known as the Five Points Community. The shopping center, named Eastgate Shopping Center, was fully leased within a short period of time and generated a relatively good income for the college. Eastgate appeared to be a good investment because the North Carolina Presbyterian Home was located just across the street. This relatively large retirement home drew residents from a broad geographic area, especially from throughout the piedmont region of North Carolina. These residents were assumed to be potential customers for the stores in Eastgate Shopping Center. This was never as successful as originally thought, and the North Carolina Presbyterian Home was relocated to Colfax, North Carolina, and renamed River Landing. The biggest reason that this was not a good investment was the general deterioration of the area and demographics, such as rising crime rates in the area.

Joining with Other Colleges and Universities in the Area to Increase Enrollment?

In an attempt to increase the headcount of students on campus, High Point College joined the Greensboro Regional Consortium for Higher Education in 1972. This consortium, which included Bennett College, Greensboro College, Guilford College, the University of North Carolina at Greensboro and North Carolina A&T University, allowed students at any of these institutions to attend courses at any of the other member institutions.

Offering Evening Degree Courses for Working Adults

Capitalizing on a void in evening educational offerings in High Point and Winston-Salem, the most successful program to increase enrollment at High Point College from the late 1970s into the 1990s was the evening degree program. The program began in 1978 as the Continuing Adult Education Program (CAEP), which was offered to working adults, many of whom had two-year degrees. Participants in this program could earn their four-year bachelor's degree in business administration, sociology, psychology and communications while working full time. Evening classes were taught in the R.J. Reynolds World Headquarters Building in nearby Winston-Salem.

The Continuing Adult Education Program was so successful that it was expanded to include more majors, and High Point College/University eventually built a satellite campus in Winston-Salem. Evening classes were also taught on the main High Point campus in areas such as interior design and furniture marketing. The success of this evening degree program was a welcome addition in a time when the number of students going to community colleges was increasing and a smaller percentage of those who were going to four-year colleges were choosing private liberal arts colleges.

New Building Construction Reflected Dr. Martinson's "Steady as You Go" Philosophy

Dr. Jacob Martinson, whose twenty-year tenure as president of High Point College began in 1985, had a number of successes, including the construction of needed buildings on the campus. Of the five major buildings that were constructed between 1985 and 2005, four—Finch Residence Hall, Hayworth Fine Arts Center, Norton Hall and David Hayworth Hall—are discussed in chapters 7 and 8 because they are named for individuals or families related to the furniture industry. The fifth major building was the James H. and Jesse E. Millis Athletic and Convocation Center. This convocation center was built around the existing university gymnasium and greatly increased its size and usability. Some

Dr. Jacob Martinson, president of High Point College/University from 1985 to 2005. *Courtesy of High Point University.*

of the changes that were made included state-of-the-art seating, modern electronic screens and a scoreboard in the athletic arena, more up-to-date locker rooms, classrooms, offices and facilities for meetings and dinners. The Millis Center also brought a needed boost to the High Point Athletic Program, and it provided additional space for large meetings like the fall convocation, the Veteran's Day celebration and the Christmas Community Prayer Breakfast.

LET'S CHANGE THE NAME FROM COLLEGE TO UNIVERSITY

High Point College became High Point University in 1991. *Courtesy of Richard Bennington.*

During the first few years of Dr. Martinson's presidency, a committee was formed to study the future of High Point College. As the committee discussed the best way to chart a forward direction for the institution, the subject of a possible name change arose. They decided that changing the institution from a college to a university would provide an umbrella under which new programs of study and other enhancements could be added. Although several other names were considered, there was widespread support for keeping the name "High Point" to reflect the heritage of the institution and its long-standing ties to the community. Therefore, on October 9, 1991, the board of trustees unanimously voted to change the institution's name from High Point College to High Point University.

THE CAMPUS WAS UNIFIED FOR SAFETY AND SECURITY

Dr. Martinson and his staff also sponsored an initiative to have Montlieu Avenue rerouted to the south side of the campus. At that time, Montlieu Avenue ran through the middle of the campus, following an east–west route in the area that is now occupied by the Gene and Jane Kester International Promenade. This meant that students had to dodge traffic on Montlieu

Avenue, also a state highway, when going from their residence halls to classes in academic buildings like the Haworth Science building or to attend chapel services. Dr. Martinson and his staff were able to work with state government officials in Raleigh to get Montlieu Avenue rerouted to what was then the outer perimeter of the campus, which greatly increased the security of the students.

THE QUBEIN YEARS:
A DRAMATIC CONTRAST IN LEADERSHIP STYLE

Dr. Jacob Martinson retired in 2005, and the High Point University Board of Trustees asked 1970 High Point University graduate and internationally prominent motivational speaker Dr. Nido Qubein to assume the presidency of the university. Dr. Qubein's background and experience when he was selected for the presidency was quite different than Dr. Martinson's background when he was chosen twenty years earlier.

When Dr. Martinson was chosen for the presidency, he was already an experienced college president, having served successfully as president of two small private colleges, Andrew College in Georgia and Brevard College in North Carolina. When he was president of Brevard College, the position he held when he was selected to be High Point's president, Brevard had been cited as one of the top-ten two-year private colleges in the country. He came to High Point University because he wanted to move from a two-year college to a four-year college. He knew how to work with a board of trustees and had developed a mindset of slow, controlled growth, such as expanding the infrastructure of the college by constructing one new building at a time. He had been successful because he knew the ropes. And he was a good president for High Point College/ University, with many notable successes during his tenure.

Dr. Nido Qubein, High Point University president from 2005 to today. *Courtesy of High Point University.*

In contrast, Dr. Qubein came with no preconceived ideas and a "can-do" attitude. This is the attitude he had when he came to the United States as a young man with fifty dollars in his pocket, unable to speak English. His remarkable life story took him to Mount Olive College in eastern North Carolina and to a summer job at a YMCA camp in the mountains of North Carolina. It was there that he met a mentor who introduced him to the city of High Point and subsequently persuaded him to transfer to High Point College.

In the twenty-five years between his graduation from High Point College and his assuming the presidency of High Point University, Dr. Qubein had become an internationally acclaimed motivational speaker and author. When he took over the presidency, he brought his well-honed speaking and writing skills, as well as a desire to help his alma mater and his adopted hometown of High Point. He had a history of talking to people about "thinking outside the box" or maybe even throwing out the box altogether. Where Dr. Martinson "knew the ropes," Dr. Qubein has been constantly redefining the ropes.

TRANSFORMATION THROUGH MARKETING AND INNOVATION

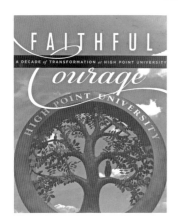

The first decade of the Qubein presidency was marked by Faithful Courage. *Courtesy of High Point University.*

Some time ago, I received a book titled *Faithful Courage: A Decade of Transformation at High Point University*, which chronicles the first ten years of Dr. Qubein's leadership at the institution (2005–15). As I read the book and thought about its title, I realized that the words *faithful courage* provide a very accurate description of Dr. Qubein's willingness to both see potential and act on it. When he took office in 2005, he saw the opportunity for the university to become more than just a good church-related, four-year liberal arts institution. He used the positive-thinking attitude that he had advocated during his motivational speeches to propel High Point University to a whole new position in the marketplace. Using an "all hands on deck" approach, the campus has experienced a complete metamorphosis of staff, programs and facilities. The following are snapshots of some of the dramatic changes that have occurred during Dr. Qubein's presidency as reported in *Faithful Courage*. (And there has been even more growth and other positive changes since the book was published in 2015.)

Eight New Academic Schools

Since 2005, eight new academic schools have been established at High Point University. They are the School of Art and Design, Qubein School of Communication, Congdon School of Health Sciences, Fred Wilson School of Pharmacy, Wanek School of Undergraduate Sciences, Plato Wilson School of Commerce and Webb School of Engineering. In addition, a new School of Dental Medicine and Oral Health has been announced and will join the other academic schools at the beginning of the fall semester in 2021. All but one of these schools are located in either newly constructed or extensively remodeled state-of-the-art facilities.

Increased Number of Faculty, Students, and Size of Operating Budget

The size of the full-time faculty tripled from 108 in 2004 to approximately 366 in 2020. Over the same period, the traditional undergraduate enrollment increased from 1,450 to 4,600. Another indicator of the dramatic growth is the size of High Point University's operating and capital budgets, which expanded from $38.0 million in 2004 to $300 million in 2020. Over the same time period, the economic impact of the university has increased from $160.3 million to $500.0 million. In addition, graduate programs, including masters in communication, business, and education were added. A doctoral program in educational leadership was also added as well, as multiple graduate degrees in the health sciences.

Table 3

The Numbers Tell the Story of HPU's Growth Under Dr. Qubein

	2004	2020	Increase
Traditional students	1,450	4,600	217%
Full-time faculty	108	366	239%
Campus size (acres)	91	500	449%
Buildings on campus	28	122	336%
Total positions	385	1742	352%
Economic impact	$160.3 million	$765 million	377%
Operating and capital budget	$38 million	$300 million	953%

Source: High Point University

Table 3: The numbers tell the story of High Point University's growth under Dr. Qubein. *Courtesy of High Point University.*

GREATLY INCREASED SIZE OF THE CAMPUS AND NUMBER OF BUILDINGS

In 2004, the campus covered ninety-one acres on the eastern side of the city of High Point. By 2020, HPU had acquired much of the adjoining land on the west, north and northeastern sides of the campus, and the physical size of the campus footprint had increased to five hundred acres. During that same period, the number of buildings increased from 28 to 124, including both new and acquired. The increased number of buildings includes a number of newly constructed buildings all across the campus. The following list of new building construction provides evidence of the dramatic changes occurring on campus.

- **Academic Buildings**: Earl N. Phillips School of Business, Nido R. Qubein School of Communications, Plato S. Wilson School of Commerce and Stout School of Education. Since 2015, the following academic buildings have been built or are being built: Congdon School of Health Sciences, Fred Wilson School of Pharmacy, Wanek School of Natural Sciences and Caine Conservatory.

- **Athletic Facilities**: Coy O. Willard Baseball Stadium, Jerry and Kitty Sports Center, Witcher Athletic Center and Vert Stadium. The Nido and Mariana Qubein Arena and Conference Center, which is currently under construction, will add new athletic facilities, as well as a venue for graduation and other large gatherings and a five-star hotel.
- **Student Housing**: Aldridge Village, Blessing Hall, Greek Village, Centennial One and Two and York Hall.
- **R.G. Wanek Center**: A complete student center, the R.G. Wanek Center contains student rooms, three eating venues, a Jamba Juice, a theater, game room, twenty-four-hour satellite library, a fine dining restaurant and a convenience store.
- **Academic Services**: Cottrell Hall contains the Career and Professional Center, Internship Center, Office of Study Abroad, Entrepreneurship Center, Harris Sales Education Center, Center for Undergraduate Research and Creative Works, Student Success Center and conference and meeting rooms. The Webb Conference Center provides meeting rooms for many different diverse groups across campus and classes that benefit from a discussion-type format.

Other buildings that have been constructed since 2015 or are being constructed include the Webb Conference Center, the Caine Conservatory and additional student housing buildings. All existing facilities that were not replaced by new construction have been updated and remodeled. These include dormitories across campus and other non-dormitory housing centers that have been acquired, Hayworth Chapel, Smith Library, university entrances, James H. and Jesse E. Millis Athletic and Convocation Center, Norcross Graduate School (formerly Cooke Hall) and Couch Hall. Two other new buildings that are in the planning stage are a building to house a new library and admissions center and a second building to house a new School of Dental Medicine and Oral Health.

CAMPUS BEAUTIFICATION PROJECTS

There have also been several campus beautification projects and other additions that have been added to make the campus especially inviting. Among them are:

- The Mariana H. Qubein Arboretum and Botanical Gardens, which includes twenty-four gardens featuring more than three thousand types of plants and four hundred types of trees.
- The David Hayworth Park with an amphitheater, sculpture garden and botanical gardens.
- The Gene and Jane Kester International Promenade, which features three fountains, metal benches (many with sculptures of famous people), flags representing the many countries from which HPU students are from and artistically designed brick sidewalks.
- Campus entrances on both the east and west sides of the campus warmly greet visitors and introduce the friendly, safe and inspiring campus.

A DEDICATION TO SERVICE LEARNING

The High Point University website states, "Service Learning is where service, leadership and ethics meet for the common good." This is a clear, concise way of expressing the mission of the service learning program at HPU, whose mission is "to engage students in an experiential and interdisciplinary learning environment that promotes their understanding of and commitment to responsible civic leadership." It is a way of connecting what is taught in the classroom with the real world and the practices of good citizenship.

Bonner Leaders and Volunteers in Service to America (Vista) are two very active programs on the High Point University Campus that get High Point University students and graduates involved in programs to help alleviate poverty in the High Point area.

Other examples of service learning classes or projects are:

- Interior design classes working on projects for local nonprofits, such as helping the YWCA, HPU Community Writing Center and West End Ministries create a more usable space for their activities.

- English majors who have tutored refugee children.
- Exercise science majors who have offered free lessons in healthy cooking and exercise at local community centers.
- Business majors who have helped nonprofits find financial stability.

DOCUMENTARY PHOTOGRAPHY
LEARNING EXPERIENCE FOR STUDENTS

An example of a service learning class is the Documentary Photography class. For one service learning assignment, the students in this class made the community their classroom by visiting local furniture manufacturers Braxton Culler, Baker, Davis and Edward Ferrell/Lewis Mittman to document

Photographs from the High Point University documentary photography class. *Courtesy of the High Point Historical Museum.*

the community through photography. They did this by photographing the individuals who were building the furniture in these High Point–area factories. They were then able to showcase their photography skills through two shows of their work at the High Point Historical Museum. One show was titled "Out of the Woodwork," and the other was called "Hands Behind the Craft." Their photographs are now part of the permanent collection of the museum.

The students were also able to learn how furniture is made, sold and distributed by talking with the workers. The students commented to Ms. Benita Van Winkle, associate professor of art and course instructor, that their experience while visiting the factories and talking with the workers also allowed them to really connect with High Point as a city and the people who live there, rather than it just being the place where they attend college.

Experiential Learning

The first milestone of the transformation of High Point University's curriculum was to stress that student learning should no longer be limited to the classroom. To quote Dr. Qubein:

> *Students must learn to apply the classroom content through experiential learning, and they must prepare themselves in a way that says, "I know what the world expects of me, and I can bring something of value." There is an old Chinese proverb that says, "I hear, and I forget. I see, and I remember. I do, and I understand."*

Dr. Qubein's administration stresses the importance of students participating in internships and other work experiences during their time at High Point University. Home furnishings, interior design and visual merchandising students, along with many other majors, have benefited greatly from various work experiences with the High Point Furniture Market and other furniture-related businesses. These experiential learning opportunities have proven to be a particularly strong point in the partnership between High Point University and the furniture industry.

Dr. Nido Qubein teaching a seminar on life skills. *Courtesy of High Point University.*

PRESIDENT'S SEMINAR ON LIFE SKILLS

Dr. Qubein teaches seminars on life skills, emphasizing the same lessons he has shared with top executives across the country. He feels that interpersonal skills, collaborative working skills and the ability to sell yourself and your ideas are essential to a successful career. He speaks to the freshman seminar twice a week in the fall semester and to seniors twice a week in the spring semester. The freshman seminar focuses on the best practices new students can utilize immediately, such as time management skills. In this class, they write papers outlining their strengths, weaknesses, goals and plans for college.

For seniors, the two-session-per-week seminar is tailored toward managing professional relationships and tackling life after college. Underlying both seminars are the ideas that interpersonal skills, collaborative working skills and the ability to sell yourself and your ideas are essential to a successful career. Dr. Qubein also discusses the "rule of thirds," which is investing one-third of one's life in earning, one-third in learning and one third in serving.

Chapter 11

VALUES AND VIRTUES

Similarities Between High Point University and the Furniture Industry

I am inspired when I hear High Point University's values and virtues explained using such terms as *generosity and gratitude*; *God, family and country*; *life skills and civility*; and *respect*. These are all worthy cornerstones on which the educational experience at the university is based. And when I read how each of these is put into practice, it is apparent that the students are exposed to the values and virtues that will help them be honest, properly grounded adults when they leave the campus. They are taught life skills that will serve them well.

My experience has convinced me that the furniture industry recognizes and puts into practice many of the same values and virtues that are being taught at High Point University. Let's first look at High Point University and then the furniture industry to see where these similarities exist.

HPU MOTTO AND EMPHASIS ON "GOD, FAMILY AND COUNTRY"

Perhaps a good place to start is High Point University's motto, which is: "At High Point University, every student receives an extraordinary education in an inspiring environment with caring people." In other words, High Point University aspires to be out of the ordinary, to inspire students to do their best and to show care and respect to everyone. Now, let's couple the institution's motto with a portion of Dr. Nido Qubein's statement from the book *Faithful Courage*, in which he is describing HPU as an entrepreneurial institution:

> *HPU stands out, in part, because our students are given ample experiential learning opportunities that extend beyond the classroom. They are also inspired by the values and principles that form the foundation for our institution, which is a God, family and country school. We distinguish ourselves by rendering results that enable every student to lead lives filled with purposeful opportunities and meaningful endeavors.*

As Dr. Qubein proudly states in all the High Point University publications, at events on campus and to anyone who asks, "HPU is a God, family and country institution." There are a number of activities and events that stand out in my mind as excellent examples of what Dr. Qubein is talking about. Among them are High Point University's chapel programs, Veteran's Day celebration and the Christmas Community Prayer Breakfast.

Hayworth Chapel and Chapel Programs

High Point University has a very active, involved religious life program centered around such activities as the Wednesday chapel services, which are held at 5:30 p.m. throughout the academic year. These programs involve preaching, singing and a variety of campus organizations, such as members of fraternities and sororities serving as ushers and other "worship assistants." The services are followed by group discussions with the campus minister, and university and other worship leaders from the preceding service.

Hayworth Chapel is also available during times of need. One instance that will always stick in my mind is September 11, 2001. I was walking down the hall near my office when I heard our departmental secretary, who was busily working at her computer, exclaim, "My goodness, airplanes have just hit the World Trade Center in New York!" Shocked and not knowing what to do, I hurried to Hayworth Chapel, which was soon overflowing with faculty, staff and students. It wasn't long before Bishop Tom Stockton, a retired Methodist bishop who, at the time, was working at HPU as a "bishop-in-residence," spoke briefly to the group. I can't quote exactly what Bishop Stockton said, but his remarks were just the right thing to give us comfort.

129

A chapel program in the Charles E. Hayworth Sr. Memorial Chapel. *Courtesy of the High Point University Chaplain's Office.*

There are other chapel programs that I have found meaningful, just one of which is the annual Martin Luther King Day program. Each year, a prominent speaker, usually a minister from the Black community, is invited to come and speak to the campus as it honors the life and work of the late social activist and minister.

Although Hayworth Chapel, with its weekly worship services and other events, reflects High Point University's Methodist heritage, the religious life program is designed to provide hospitality for people of all traditions. HPU's Catholic campus ministry hosts Catholic mass on Sunday evenings, HPU's Hillel hosts celebrations for all major Jewish holidays, and Muslim, Hindu, Buddhist and students of other faiths use the multifaith prayer and meditation space in the chapel. The Interfaith Dinner Club, which meets monthly, and the Angel Tree, which buys Christmas gifts for children in the community, are two ways that the religious life program encourages spiritual growth among students.

The High Point University Choir sings at the annual Christmas Community Prayer Breakfast. *Courtesy of High Point University.*

CHRISTMAS COMMUNITY PRAYER BREAKFAST

For almost fifty years, near the end of each fall semester, High Point University has invited the greater High Point faith community to the campus for a community prayer breakfast. I, like a lot of other High Point area residents, look forward to this annual event as a means of touching base with friends and setting my mind clearly on the true meaning of Christmas. Each year, there is an inspiring message by a nationally prominent minister and special music provided by the High Point University Chapel Choir.

VETERAN'S DAY CELEBRATION

Each year, in recognition of patriotism, one of the core values of High Point University, veterans are invited to the campus for a patriotic program and breakfast to honor the sacrifice and service of the men and women who have served in the United States military. The celebration consists of a complimentary breakfast, a presentation by a notable speaker and a selection

High Point University's annual Veteran's Day program. *Courtesy of High Point University.*

of patriotic musical performances. Hundreds of veterans from the Piedmont Triad area look forward to attending this meaningful program.

Although I am not a veteran, I had heard so much about this celebration that I volunteered to help with the event. I was amazed at the diversity of those who attended. There were veterans of World War II, the Korean War and the Vietnam War, as well as those who have served around the world in various other conflicts. A few even brought their service dogs. It was a truly moving experience to see the reactions of all these veterans as the band played "God Bless America" and other patriotic songs.

THE HIGH POINT UNIVERSITY COMMUNITY CHRISTMAS CELEBRATION

Beginning in 2011, High Point University began an annual tradition known by most people as simply Community Christmas. This tradition, which reinforces the fact that High Point University is a God, family and country institution, is where the entire High Point community is invited to bring their families and enjoy the spirit of Christmas. This event, which has brought as many as twenty-five thousand people to the campus annually, takes place each December, just after the students have gone home for Christmas break.

Visitors park in an off-campus lot and ride buses to the campus, which is transformed into what I call the "world of Christmas." It is staffed by High Point University volunteers who help make the community visitors' trip to the campus a magical experience. They can enjoy everything from a life-size

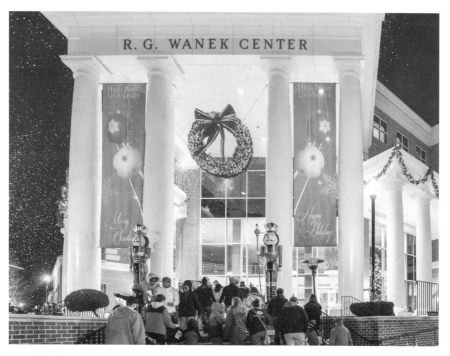

The High Point University campus turned into a world of Christmas. *Courtesy of High Point University.*

nativity scene and a Christmas program in the chapel, to children having a chance to sit on Santa's lap and tell him what they want for Christmas. There are carriage rides; a petting zoo; a hot dog and potato chips supper; a snow-making machine on the top of the Wanek Center, filling the night sky with white fluffy "snow"; and hot chocolate from food carts. This High Point University Community Christmas has become an annual event for many families in the area.

VALUES AND VIRTUES OF THE FURNITURE INDUSTRY

As I mentioned earlier, when I think back over my many years of studying the furniture industry and associating with furniture professionals, I have always thought that many of the same values and virtues were being practiced by both High Point University and the furniture industry. This was validated by an internet survey conducted by the leading industry newspaper, *Furniture Today*, in 2005. The survey was directed to retailers, manufacturers, and suppliers in the home furnishings industry. These industry professionals were asked the question: "What skills should be taught by colleges and universities to prepare young people for a career in the home furnishings industry?" The intent of the survey was to help ensure that what was being taught in the various home furnishings and interior design classes were on-target with industry needs.

When the results were counted, the number-one skill or area of knowledge desired by the executives responding to the survey was business ethics, which seems very compatible with the values and virtues of High Point University. (Business ethics was followed closely by a desire for new hires to have a knowledge of consumer behavior, also something taught at High Point University). The remainder of the top eight skills or pieces of knowledge desired by the executives were visual display/visual merchandising, teambuilding, components of merchandising, retail math, advertising principles and public relations skills.

It is significant, however, that on this national survey, which was a relatively even representation of all sections of the United States, ethics

General Colin Powell addresses the American Furniture Hall of Fame Induction Banquet. *Courtesy of American Furniture Hall of Fame.*

was overwhelmingly rated as the top skill or attribute that should be taught to young people who want a home furnishings industry–related career. In fact, on a scale from 1 to 5 (with 1 being not important and 5 being very important), the average score for "ethics education" was 4.7.

AMERICAN FURNITURE HALL OF FAME

This connection of similar values and virtues between HPU and the furniture industry became even clearer in the fall of 2018, during the thirtieth annual induction ceremony of new members into the American Furniture Hall of Fame. As one of approximately 1,300 attendees listening to acceptance remarks by the four inductees, I was impressed by the fact that they were passionate about putting into practice the values and virtues that the faculty and staff of High Point University are trying to instill in HPU students. To quote some excerpts from the inductee acceptance speeches:

> *To future leaders: while I encourage you to continue to lead in key areas like innovation, design, quality and customer experience, never forget the impact that you make on someone's life. Together, we transform houses into homes.*
> *—Kurt L. Darrow, chairman, president and CEO of furniture manufacturer La-Z-Boy Inc.*

> *We bring innovation to another level and another speed. When you out-innovate the marketplace, you have the chance to supply the consumer with what they want.*
> *—Allen E. Gant Jr., chairman and recently retired CEO of fabric supplier Glen Raven Inc.*

> *We continue to look at our business from our customer's point of view and try to innovate ways to make her shopping experience better. Furniture shopping needs to be fun.*
> *—Keith Koenig of South Florida retailer, City Furniture*

> *Our vision is to enhance the lives of the people we touch.*
> *—Paul Toms Jr., chairman and CEO of furniture manufacturer Hooker Furniture Co.*

These four inductees joined an elite group of 115 other hall of fame members. The list of the names of the other members reads like a who's who of the home furnishings industry. The selection of individuals to be inducted into the American Furniture Hall of Fame is based on four criteria: enduring excellence, superior accomplishments, innovation and creativity and philanthropic generosity. Again, all of these criteria are very compatible with the values and virtues of High Point University.

NORTON'S ADVICE TO THE STUDENTS

During my years of teaching, I have found professionals in the furniture industry more than willing to speak to classes or other groups of High Point University students about their philosophy of life and work. There are several I could bring out as outstanding examples of industry professionals sharing their helpful advice with the High Point University students, but one example of advice that was particularly well received was a presentation made on April 4, 2008, by the late Patrick H. Norton, a member of the American Furniture Hall of Fame and former chairman of La-Z-Boy Inc. They seemed particularly responsive to his remarks since he was speaking to them in the lecture hall of the building on the High Point University campus that bears his name.

These are some of the points I took away from Norton's presentation:

"What I look for when hiring new employees."

- *Integrity (personal integrity)*
- *Enthusiasm (genuine enthusiasm about themselves and the industry)*
- *Someone who is not a "clock watcher"*
- *Dedication (someone who will dedicate themselves to what they are doing because it will make tomorrow better)*
- *Desire to learn (someone who is inquisitive and wants to learn)*

Norton's further advice to the students:

- *Don't let money be the driver! Get a job you like and where you can learn and progress in your career. Get in a position where people like you and will help you.*

Mr. Pat Norton giving advice to High Point University students in Norton Hall. *Courtesy of High Point University.*

- *Network. Start early and get to know people. Find a mentor—that person can help you greatly.*
- *Distinguish yourself with grades, by working hard on the job.*

FURNITURE FELLOWSHIP PRAYER BREAKFAST

When I think of God, family and country, one of the first things I think of is the Furniture Fellowship Prayer Breakfast, where the entire furniture industry is invited to have breakfast, hear an inspirational message and inspirational music. (In many ways, it reminds of the Christmas Community Prayer Breakfast at High Point University). Modeled after the congressional prayer breakfast in Washington, D.C., the Furniture Fellowship Prayer Breakfast is held in High Point during the spring furniture market. Attendees of the prayer breakfast regularly fill most of the seats in a large banquet hall. For this particular year's breakfast (shown in the image on page 138) the speaker was furniture industry executive Chad Spencer.

Furniture industry executive Chad Spencer addressing the Furniture Fellowship Prayer Breakfast. *Courtesy of Furniture Fellowship.*

I think it speaks a lot about the values and virtues of the individuals in the furniture industry that such a large number of them would take time at the beginning of one of the busiest days of the market to pause, reflect and listen to an outstanding inspirational speaker. Past speakers at the Furniture Fellowship Prayer Breakfast include Dr. Nido Qubein, High Point University president; Dr. Gary Chapman, prominent minister and *New York Times* bestselling author; Kathy Ireland, former model, actress and current president of Kathy Ireland Worldwide; and Dan Cathy, CEO of Chick-Fil-A.

THE CULP FAMILY AND CULP INC.: AN EXAMPLE OF ETHICS, DEDICATION AND SERVICE

Although there are many companies in the home furnishings industry that can be held up as having high ethical standards, Culp Inc. and the

Culp family is an outstanding example of high ethical standards, along with dedication and service to High Point College/University. Culp Inc. was founded in 1972 by Robert Culp Jr. as a wholesaler of fabric for the upholstered furniture industry. Over the years, the company has expanded to the point that, today, it is one of the world's largest manufacturers and distributors of mattress fabrics and fabrics for upholstered furniture.

The partnership between Culp Inc. and the Culp family and High Point College began in 1990, when Robert Culp and his wife, Esther, established the Culp Chair of Ethics. The objective of the chair of ethics was to provide a fund from which the income could be used by High Point College/ University to employ faculty to teach courses in ethics and values.

As I thought about the Culp Chair of Ethics and heard more about Robert Culp, who was called "Bullet" by his friends, I found myself wanting to know more about this outstanding gentleman. When I recently read a short history about Culp Inc., I discovered that Robert Culp's nickname came from the fact that he was quick to get to the point, fast on his feet, sharp-witted and moved at high velocity. In addition to those attributes, a number of other characteristics were listed, such as inspiring, charitable, honest, leader, can-do, generous, family man and civic leader. All of these are values and virtues emphasized by High Point University.

To quote the Culp Inc. history: "Learning by example, employees at Culp, from the early years until the present day, become 'Culpmatized.' That means embracing a culture of being ethical, respectful and hardworking with a desire to do 'whatever it takes' to delight customers. Most of all, 'Culpmatized' associates always find a way!" Culp Inc. and the Culp family has "found a way" to be a very valuable partner with High Point University.

Robert Culp Jr., the founder of Culp Inc. *Courtesy of Culp Inc.*

Beginning with the Culp Chair of Ethics, which was changed in 2012 to support the High Point University director of service learning with its emphasis on ethics and commitment to responsible civic leadership, the Culp family has been extremely generous and helpful in many other ways. For example, when High Point University was raising funds for the construction of Norton Hall, Culp Inc. made a generous contribution, resulting in the top floor of the building being named the Culp Inc. floor.

Robert Culp Jr. and his son Robert Culp III, who succeeded him as the CEO of Culp Inc. *Courtesy of Culp Inc.*

Robert Culp IV, the current CEO of Culp Inc. *Courtesy of Culp Inc.*

Culp Inc. has been extremely helpful with the home furnishings and interior design classes, as it hosts class field trips to its manufacturing facilities in the High Point area. For example, it has invited the Fundamentals of Furniture class to its knitting mill, where the students can see state-of-the-art machines knitting decorative, plush fabric for mattress covers.

It has also been generous in inviting classes to its corporate offices in High Point, where its fabric designers have made presentations to the students on fabric and fabric merchandising. The students get to see the latest upholstery fabrics and are exposed to a professional presentation by the designers, who treat the students like their customers. This way, students get a glimpse of what the buyers are looking for when selecting fabrics, especially combinations of fabrics, for their products. Another example of Culp Inc.'s assistance to the students has been when one of its top salespeople has come to the campus to make presentations on effective selling techniques to the university's sales classes.

Culp Inc. is also an example of an ethical business and a successful family business. The second generation of the Culp family, Robert Culp's son Robert Culp III (or Rob) took over the management of the company from his dad. Now, with the passing of Rob Culp, his son Robert Culp IV (referred to as IV) has taken over the management of the company that was started by his grandfather, keeping intact a family tradition of running a company that is known for ethics, dedication and service.

Ron Wanek's Sculpture: A Unique Example of Patriotism and Generosity

During the spring 2007 furniture market, I was privileged to be in the Ashley Furniture Showroom to witness the unveiling of three statues of furniture industry leaders who were members of "the greatest generation"—those who fought and won World War II. The presentation of these statues, designed by Ron Wanek, chairman of Ashley Furniture Industries, is a unique example of patriotism and generosity, values important to High Point University.

The three members of the greatest generation who were honored by the statues were Pat Norton of La-Z-Boy Inc., Louis Blumkin of Nebraska Furniture Mart and Howard Miskelly of Mississippi-based Miskelly Furniture. Norton won the Distinguished Flying Cross for his accomplishments in the army air corps in the Pacific. Blumkin and

A statue of Patrick H. Norton as he would have looked in World War II. *Courtesy of Richard Bennington.*

Miskelly both fought their way across Europe. Miskelly ended the war at the Elbe River, where the Allies met the Russians, and Blumkin helped liberate the prisoners at Dachau Concentration Camp.

The statues were designed to depict these World War II veterans as they would have looked during the war. Norton's statue is on permanent display on the High Point University campus, near Norton Hall. But Wanek's deep sense of patriotism did not stop with honoring these three World War II heroes. It is even more evident in the Soldiers Walk at Memorial Park in Arcadia, Wisconsin, a fifty-four-acre memorial park that contains a 540-meter-long walk, with statues of heroes from all of the American conflicts that were donated and sculpted by Wanek. As the chief designer and benefactor of the walk, his goal was to honor these veterans for their service to the United States. Both his honoring of the three World War II veterans from the furniture industry and veterans of the area around Arcadia are evidence that his values are in-line with the values and virtues of High Point University.

Chapter 12

HIGH POINT UNIVERSITY PARTNERING WITH THE FURNITURE MARKET

I n 1979, when High Point College began offering the furniture marketing degree, it entered a new era of increased partnership with the High Point Furniture Market and the furniture industry. Both the furniture market and individual companies within the furniture industry enthusiastically embraced the idea of home furnishings–related courses at High Point College and provided the assistance needed to help ensure the success of the furniture marketing program and later the interior design and visual merchandising design programs.

There have been a variety of ways that the furniture market showroom buildings, individual companies that show at the market and the High Point Market Authority have partnered with High Point College/University. The following are examples of these partnerships:

INFORMATIONAL BOOTH IN THE SOUTHERN FURNITURE EXPOSITION BUILDING

For several years in the 1980s and early 1990s, the management of the Southern Furniture Exposition Building allowed High Point College to have a complimentary booth to spread awareness of the furniture marketing program to individuals visiting the market. The first location was on the third floor of the building, near the accessory showrooms. Later, the High Point College

The plaza outside of the International Home Furnishings Center during a spring High Point Furniture Market. *Courtesy of High Point Market Authority.*

booth was moved to the ninth floor of the building, where the college shared an area with the Bienenstock Furniture Library, which was selling books to market attendees. The traffic was much better in the second location, but both were extremely helpful for getting the word out about the home furnishings–related educational programs at High Point College/University.

DESIGN ASSISTANCE AND PROJECTS AROUND THE MARKET

In the days and weeks leading up to market, some exhibitors hire students, especially interior design and visual merchandising majors, to help them implement their showroom designs or assist their showroom designers in getting the showrooms ready for the buyers attending the furniture market. Small companies that are new to the market have also allowed High Point University interior design classes to either design their space or a designated section of their space as class projects, thereby bringing real-world experience into the classroom.

High Point University interior design students have also participated in design projects for the furniture market buildings. An example of one of these projects involved the students being given a designated space in the International Home Furnishings Center (IHFC) and asked to design it as if it were a lounge space for interior designers visiting the High Point Furniture Market. It was to be a place with wi-fi capabilities, where interior designers could work during off-market times. Specific areas for the students included in the lounge were places where relatively small seminars for designers could be held, a small library for manufacturer's catalogs and a private area where the designers could relax. The students organized the space and chose the colors, paint, carpet and light fixtures. They drew the plan, then supervised the implementation of it.

For another project, the HPU students were given a portion of the temporary trade show space in the Suites at Market Square, where eco-friendly products could be featured. The students were given the specifications, then they created the design of the display spaces. It was an educational space where manufacturers from all over the market could display specific eco-friendly pieces from their product lines. The intent of the space was to bring awareness to the eco-friendly products that were being offered by market exhibitors.

MARKET SHOWROOM ASSISTANTS

A large number of exhibitors hire High Point University students to work during the market in their showrooms. A common practice is to hire the students to act as receptionists at the front desk of the showroom. The job of these students is to greet the retailers, designers or other visitors to the space, answer questions and page the appropriate sales representative who will actually take the retailers through the space. These assistants can also act as "floaters," a situation in which they may guide market visitors through the showroom when no one else is available.

Other exhibitors may assign other duties to the students. For example, one company that manufactures higher-priced upholstered furniture hires students to assist their representatives in making presentations of the decorative fabrics they are introducing that market. Following the presentation, the student will "tidy up" that area of the showroom by refolding the fabric samples and rehanging them on the display racks so that the area will be ready when the next customer comes in.

High Point University students Sean Bando (*left*) and John Kenney (*right*) working at the reception desk of the Universal Furniture Market Showroom. *Courtesy of Richard Bennington.*

Another relatively common practice is for showrooms to have a "canteen" or area where refreshments are served. Students are sometimes hired to "run the canteen." Showrooms that serve lunches typically have a caterer who takes care of the preparation and serving of the meals.

ASSISTING THE HIGH POINT MARKET AUTHORITY

The High Point Market Authority has enlisted the help of High Point University students in a variety of ways, one of which is to hire students as "market scouts." These students are typically majoring in one of the home furnishings–related programs. These "scouts" are first trained in the overall layout and organization of the market. They then attend the First-Time Buyer's Orientation Meeting. Afterward, they meet with buyers who haven't been to the High Point Market before and help them get organized so they can make the most out of their visit. They direct the first-time attendees to the showrooms they want to visit, or they may help them find (scout out) showrooms that sell the type of products they are looking to buy, hence the name "market scouts."

Another way the Market Authority has used students is to have them assist with the educational seminars it sponsors for designers each market. This is especially valuable for HPU students who are either majoring or minoring in event management because it gives them real-world experience with organizing and conducting an event. They take care of a variety of necessary tasks, including seating the participants and scanning their badges. A third way the Market Authority uses students is to have them help set up and oversee the Market Media Center, giving them experience in operating of a real-world press center.

ASSISTANTS FOR MULTI-LINE SALES REPRESENTATIVES

Another common practice in the furniture industry is for manufacturers to use independent "multi-line" sales representatives to sell their products. This is a situation in which an independent manufacturer sales representative contracts with a number of different furniture manufacturers, or lines, to represent them in a designated sales territory. For example, a sales representative might contract with one company that sells bedroom furniture, a second that sells occasional tables and a third that sells outdoor (summer and casual) furniture. This allows the salesperson to offer a number of different categories of furniture products (hopefully in the same relative price range) to retail store owners or buyers throughout their sales territory.

This arrangement works reasonably well, except at the furniture market. The reason for this is that the sales representative can't be in "more than one

A High Point University student helping a sales representative during a High Point furniture market. *Courtesy of High Point University.*

place at the same time." To help solve this problem, a significant number of multi-line representatives have hired High Point University students to work for them in one of their showrooms at the High Point Market. A typical scenario is that the manufacturer's representative will take the students to the premarket sales meeting of the company they will be working for. By attending this meeting, the students can learn about the new product introductions the company is presenting at that market. When market opens, the student can meet dealers from the sales representative's territory and show them the new product introductions in one space while the representative is working with their dealers in another space. The sales representative is only a cell phone call away, so they can answer questions remotely or, perhaps, walk over from the other showroom.

This is a good learning experience for the students because they not only learn about the products in a given showroom, but they also learn how to greet people and make sales presentations. These sales assistants can be from various majors; the main ingredients for these positions are that the candidates need to be outgoing and approachable. Either the sales representative or someone else in the showroom will provide them with the product and company knowledge they need to work effectively in the showroom.

CLASS FIELDTRIPS TO MARKET SHOWROOMS

The High Point Market Authority allows students from High Point University, as well as other colleges and universities with home furnishings–related majors, to visit the market showrooms on a designated day, often near the end of the market, so they can see the latest in home fashions and products. Some companies partner with High Point University to have showroom tours with presentations for particular classes or groups of students.

Some especially valuable tours of a showroom at the market have been tours of the La-Z-Boy showroom, conducted by 1989 High Point graduate Eric Gatton, a sales professional who holds the Certified Home Furnishings Representative (CHR) designation and is the La-Z-Boy regional vice-president of sales for the southern central region. Eric began his High Point University career as a business major, but before graduation, he also added a major in furniture marketing.

Eric Gatton, a High Point University graduate and La-Z-Boy regional vice-president, explains a marketing program that uses the actress Kristen Bell as one of their brand ambassadors while taking High Point University students on a fieldtrip through the La-Z-Boy Inc. showroom at the High Point Furniture Market. *Courtesy of Richard Bennington.*

These tours have been valuable to High Point University students who are learning about the home furnishings industry because they involve both a detailed discussion of the features and benefits of the new product introductions and an overview of the La-Z-Boy marketing and merchandising program. For example, at the October 2018 market, La-Z-Boy introduced a new brand ambassador, actress Kristen Bell. Their brand ambassador is the face of La-Z-Boy to many members of the buying public and is a key component in their marketing program. Kristen was chosen because she is someone who could support the brand and what La-Z-Boy stands for as a company—their values, heritage and products. (A former La-Z-Boy brand ambassador is actress Brooke Shields.)

Eric's discussion of the other parts of the marketing and merchandising program shows how they all come together to help dealers to make an impact in their markets. This involves a discussion of television advertising, email campaigns, social media posts and in-store displays. Also included are schedules of suggested times for the retailer to advertise merchandise in their respective areas and premiums that may be available. In other words, how everything works together to sell the product.

Another example of a market tour leader is designer Joe Ruggiero, who has made presentations of the merchandise in his licensed collection in the Miles Talbott Showroom to High Point University interior design students. Since some of his products feature Sunbrella fabrics, the presentation also includes comments from a sales representative from Glen Raven Mills, the manufacturer of Sunbrella Fabrics.

PARTNERING WITH COMPANIES SHOWING AT MARKET

Companies with market showrooms at the High Point Market have also partnered with High Point University students to design products to be introduced at the market. An example is Shermag, a Canadian furniture manufacturer that sponsored a competition with High Point University students to design a crib and changing table to sell in the United States market. The company chose High Point University because of its ties to the furniture market and the reputation of HPU's home furnishings and design programs. The winner, Aria Real, won a cash prize and a trip to the Shermag factory in Montreal to show her designs to company officials.

High Point University student Aria Real, the winner of the Shermag Design Competition. *Courtesy of Cathi Nowicki.*

Chapter 13

ONGOING PARTNERSHIPS
WITH THE FURNITURE INDUSTRY
OUTSIDE OF MARKET

The following are some ways in which different companies or organizations within the furniture industry partner with High Point College/University other than at the furniture market.

INTERNSHIPS

A number of companies in the High Point area have let students with various home furnishings–related majors earn academic internship credit by working part time for a semester or summer session. These companies are representative of most segments of the home furnishings industry from manufacturing and retailing to design, and the companies that supply products or services to the industry. Students who have participated in these internships have been almost unanimous in their feeling that internships have been very helpful in providing them with the skills and industry insight to help them achieve success in their jobs following graduation.

An excellent example of a student benefiting from such an internship is 2017 interior design major and graduate Emily (Springer) Yavorsky. Based on the experience she gained while working during the furniture markets in High Point, Emily decided she wanted to learn more about the real world of design through an in-depth internship. She was able to secure a summer internship between her junior and senior years at Barbour Spangle Design

High Point University intern and interior design major Emily (Springer) Yarvorsky getting real-world information from Christy Barbour, a cofounder and partner of Barbour Spangle Design Group. *Courtesy of Richard Bennington.*

Group in High Point. At Barbour Spangle, she learned about commercial, residential and showroom design. Emily feels that her real-life experience at Barbour Spangle helped her during her senior year with her required design class projects.

After her summer internship, Barbour Spangle liked Emily so well that they hired her to work part time during her senior year. Through her summer internship and resulting part-time work, she gained invaluable experience, as she saw senior employees interact with real customers. She learned about the latest in materials by sitting in on manufacturer updates with the other designers.

After graduation, the management of Barbour Spangle created a full-time design assistant position so that Emily could stay with the firm, and today, she is a valuable member of the Barbour Spangle design team. Emily is convinced that her internship experience would have helped her obtain employment after graduation even if there had not been an opportunity with Barbour Spangle. Today, Emily has the NCIDQ (National Council for Interior Design Qualification) professional certification, and she thinks that working under NCIDQ-certified designers during her internship helped her do well on the NCIDQ examination.

Christi Barbour, cofounder and partner of Barbour Spangle Design, is so enthusiastic about working with Emily and the High Point University Internship Program that she submitted the following statement:

> *At Barbour Spangle Design, we strongly believe internships are an excellent way to provide students with real-world experience while exploring their career path in the design field.*
>
> *We have found that the internships we provide lead to extraordinary relationships with students like Emily Springer Yavorsky. Emily came to us as an HPU student and, upon completing her internship, was hired part-time through the balance of her senior year. During her internship and part-time tenure, she was exposed to many ideas, concepts, people and resources, giving her valuable experience and allowing her to make countless connections. In return, she brought the latest in the theory of design, as well as knowledge and an incredible attitude. The beauty of having this time with Emily meant we had time to evaluate her talent, develop a relationship with her and see the type of employee she would be. She conducted herself with grace, integrity, enthusiasm and creativity. For this reason, we offered her a full-time position before graduation.*
>
> *Building relationships is critical to advancing a career, and the experience of an internship provides an excellent foundation for learning how to develop and foster great relationships.*

PARTNERING WITH FURNITURELAND SOUTH

Furnitureland South, the world's largest home furnishings retail store, is located in Jamestown, North Carolina, just a short drive from the High Point University campus. Because of its size and the fact that it has all of its operations on its Jamestown campus, Furnitureland South has proved to be a valuable partner in offering a relevant home furnishings education. The following are examples of how Furnitureland South has partnered with High Point University:

- **Retail Sales Techniques**: Field trips to Furnitureland South and presentations by their sales management and sales training staff have been beneficial learning experiences. The staff has made presentations to classes, explaining how they take consumers through the steps to a sale. This gives the students real-world knowledge as to what it takes to successfully sell home furnishings products to today's consumers.

- **Retail Display Techniques**: With over 1.3 million square feet and over five hundred brands of home furnishings being offered for sale, visits to Furnitureland South allow students to observe a wide variety of retail display techniques. This provides an unusual look at what it takes to visually engage customers and increase the likelihood that they will purchase home furnishings products from the store.

- **Customer Service, Warehousing and Delivery**: Furniture retailing classes get a behind-the-scenes look at Furnitureland's modern warehouse, complete with the latest inventory control, storage, handling, delivery and customer service systems. This allows the students to observe what it takes to receive merchandise from vendors and place it into inventory; and when an order is received, they observe what it takes to retrieve it, perform any needed repairs or servicing, stage it for delivery and load it onto trucks in the proper order so delivery can be as efficient as possible for the company and completely carefree as possible for the customer.

- **Product Knowledge**: Classes visiting Furnitureland South get a look at many different types of products and see how they are constructed. A particularly interesting example is that students visiting Furnitureland's Sleepland mattress department learned the latest in mattress construction, as well as the benefits of using the computerized sleep diagnostic system in the mattress department, which helps ensure that consumers purchase the sleep set that is right for them.

- **Critique of Student Projects and Presentations**: Several of the Furnitureland staff have come to the campus to observe student oral presentations. They then offered their comments as a critique of the students' oral presentations. (They are not there to participate in the grading but to provide input helpful to the students' presentation techniques.)

Professor Emily Reynolds and her second-year interior design class on a fieldtrip to Furnitureland South. *Courtesy of Emily Reynolds.*

PARTNERING WITH HOME MERIDIAN INTERNATIONAL TO ALLOW STUDENTS TO GAIN UP-TO-DATE MERCHANDISING KNOWLEDGE

Home Meridian International, a division of Hooker Furniture, regularly assists in bringing realism to the home furnishings students by inviting home furnishings related classes to their corporate headquarters in High Point to see the entire product development process. This exposes the students to a real-life case study of how a company analyzes the market competition and determines which products have the best chance of success with their targeted customers.

A particularly interesting example of product development by Home Meridian is the "Highway to Home" furniture collection. The Home Meridian staff made a presentation to High Point University students on how this collection was developed for their Pulaski product line. This collection was inspired by the eclectic music and lifestyle of country music star Eric Church. To quote the Eric Church website, "With a foundation of family values, love of nature and a dash of attitude, Highway to Home is simply music for the home." The Home Meridian staff explained how the various pieces were developed to reflect the theme and how it was carried out through the various pieces in the collection.

Later in the semester, the students visited the Home Meridian Showroom at the High Point Market so they could see the actual products and how they

Occasional furniture pieces from Home Meridian International's Eric Church Highway to Home Collection. *Courtesy of Home Meridian International.*

were presented to potential buyers. The presentation of the "Highway to Home" collection began with the class being treated like any other market visitors to their showroom. They were first ushered into a specially designed sound room, where they were shown a video of Eric Church performing on stage. Then the doors were opened, and they walked out into the section of the showroom where the "Highway to Home" collection was displayed. In other words, the visitors were introduced to Eric Church first, then the collection bearing his name. It was interesting to see the detail of each piece reflecting the country music "Highway to Home" theme, even down to the hardware on the case pieces being designed to resemble parts of Eric Church's guitar.

PARTNERING WITH PROFESSIONAL INDUSTRY-RELATED ASSOCIATIONS

For most of the last forty years, High Point College/University has worked with professional associations in the furniture and design industries in various ways. Some have held seminars or workshops; others have provided scholarships or awards based on project-related contests or other competitions. The following is an overview of how the College/University has partnered with a variety of professional associations.

National Home Furnishings Association (NHFA): Now the North American Home Furnishings Association

In the late 1980s and early 1990s, the National Home Furnishings Association held its annual Management Development Institutes on campus. The institutes were organized like a one-week boot camp, with the participants staying in the college's dormitories, attending classes in the student union and eating in the college's cafeteria. The participants in the institutes were the up-and-coming management personnel from NHFA-member single-market or chain stores located throughout the United States and Canada.

High Point College provided the facilities, but NHFA brought in the faculty to conduct sessions on a variety of subjects that were helpful to retail store managers or potential managers. A successful retailer would provide a keynote speech to set the stage for the week-long institute, but then the sessions would focus on individual areas of concern, such as accounting/finance, store planning, sales and sales training, advertising/marketing and logistics/warehouse management.

Two examples of sessions that were conducted at these Management Development Institutes are particularly memorable. Neil Forney, a partner in Bryant Forney Store Planners in San Francisco, California, conducted sessions on how to organize and lay out an efficient, productive retail furniture store that will help in the sale of the product. He covered how to design a retail store in a way that is welcoming, easy to shop and productive from the standpoint of sales per square foot.

Cathy Manley of Cathy Manley Associates in Rochester, New York, covered the psychology of selling and how a well-trained salesperson can move customers through the various steps in sales from the greeting all the way from overcoming objections to closing the sale. She also covered other necessary sales-related topics, such as the importance of having effective sales meetings and strategies to increase sales productivity.

International Home Furnishings Representatives Association (IHFRA) CHR Programs

In the mid- to late 1990s, High Point University was host to the IHFRA Certified Home Representative (CHR) training program. Attaining the CHR designation was thought to be a way of increasing the professionalism of the independent sales representative. Using a schedule similar to the NHFA Management Development Institute, the CHR program was conducted by outside experts who provided the training. Some of the class modules taught attendees how to manage a sales territory, how a manufacturer's representative can work with retail furniture dealers to improve the profitability of their bottom line and improve their communication skills. Sitting in on these sessions really solidified and expanded the knowledge I was first exposed to when I rode with the Bassett sales representative during the spring of 1979, especially how a professional sales representative can wear many hats and partner with their customers.

At the end of the program, the participants were given an extensive take-home exam to complete and send back. Those who successfully completed the exam were awarded the Certified Home Furnishings Representative (CHR) Professional designation. In 1997, I paid my fee and was allowed to enroll in the training program just like all the sales reps. I spent approximately twenty hours on the take-home exam, but I passed and was one of the thirty individuals to be awarded the CHR designation at the fall 1997 Furniture Market.

International Society of Furniture Designers (ISFD) (Formerly American Society of Furniture Designers [ASFD])

In the mid-1990s, the university was approached by the American Society of Furniture Designers about partnering with them to energize their organization and their Pinnacle Design Award. Fortunately, I was able to help them find a new, professional executive director and assist them in formulating a plan to rework the Pinnacle Awards. The Pinnacle Design Awards are now held annually to recognize the best home furnishings products on the market in a variety of categories. Examples of the product categories for which Pinnacle Awards are presented are living room upholstery, summer and outdoor, formal dining, bedroom, casual dining and home office furniture. In 1997, I was proud to receive a lifetime honorary membership in the organization and am happy that, based on our combined efforts, the Pinnacle has become a prestigious design award. For several years, the Pinnacle Award judging was held on the HPU campus, and now, the university has a standing representative on the panel of judges.

In 2018, the American Society of Furniture Designers was rebranded as the International Society of Furniture Designers (ISFD), answering a growing call for an international organization devoted to advancing, improving and supporting the practice of furniture design.

International Furnishings and Design Association (IFDA)

The Carolinas chapter of the International Furnishings and Design Association (IFDA) has an awards banquet annually to honor the rising stars in design from colleges and universities across North and South Carolina. The rising stars are students chosen by their respective institutions as their most promising young talent, those who will make positive and important contributions in their future careers. High Point University has been chosen to be part of this program for the last several years. Each student is invited to choose one recent project that best exemplifies their work and design philosophy. These projects are displayed at a special cocktail reception held in their honor before the awards dinner.

Kendra Purdie and Marina Berenguer, 2018 High Point University IFDA stars. *Courtesy of Richard Bennington.*

Withit: Women in the Home Industries Today

Withit was founded in 1997 with the mission: "To encourage and develop leadership, mentoring, education and networking opportunities for professional women in the home and furnishings industries." This mission has led Withit to partner with High Point University in a variety of ways.

First, a number of HPU female students who are majoring in an area that would logically lead them to careers in the home furnishings industry have been recipients of Withit Scholarships. Along with the financial assistance provided by these scholarships, the students receive student memberships in Withit, and some were given additional scholarships to attend national Withit conferences, where they were able to view presentations on a variety of relevant subjects, many given by prominent industry speakers. Other students have been able to attend these national conferences on a reduced-fee basis.

Student memberships are also available to students with other majors that would logically lead to a career in the home furnishings industry. A large number of High Point University students participate in the Withit Day of Mentoring at the High Point Furniture Market. Participation in this day of mentoring allows students to get to know professional Withit members, attend a variety of seminars and go on field trips throughout the market. Attendance at both the national conferences and the Day of Mentoring are often seen by the students as valuable learning experiences.

International Interior Design Association (IIDA)

In 2015, largely through the efforts of HPU professor and IIDA Board of Directors member Cathy Nowicki, High Point University was established as the only private university or college in North or South Carolina to have a campus center (a campus center is the IIDA name for a student chapter). The HPU Campus Center is affiliated with the IIDA Carolinas Triad City Center, which has professional members in Greensboro, Winston-Salem and High Point. The International Interior Design Association Campus Center at HPU was considered to be a natural fit because IIDA is committed to enhancing quality of life through excellence in interior design and advancing interior design education.

To cite an example of the good fit between IIDA and High Point University, in 2017, the High Point University IFDA Campus Center received a contribution from the IIDA Carolinas Triad City Center to

(Right to left) High Point University professor and IIDA faculty advisor Cathy Nowicki; IIDA student members Lauren Yoder and Maureen Coleman; and Destiny House manager Dana Bentley. *Courtesy of High Point University.*

allow High Point University interior design students to refurbish Leslie's House, an emergency shelter for women in High Point. The students were able to give Leslie's House a much-needed facelift, just in time for the shelter's tenth anniversary.

They felt it was a valuable learning experience to see their designs come to life in a real setting and to see how they helped to make Leslie's House warmer and more inviting for the women served by the emergency shelter. The refurbishing of Leslie's House is also a great example of how High Point University is serving the city of High Point.

PARTNERING WITH THE HIGH POINT–AREA FURNITURE AND DESIGN COMMUNITY

In 2010, High Point University decided to take some funds from the Knabusch-Shoemaker International School of Home Furnishings and Design to sponsor a series of seminars to help the High Point–area furniture and design community better understand what was happening in the emerging world of internet marketing. Individuals who attended the sessions were given knowledge that would hopefully allow them to take advantage of the possibilities offered through internet marketing and help them keep up with the competition.

Seminars were offered in 2010, 2011 and 2012, featuring speakers and panel members from successful internet marketing companies and other organizations throughout the United States. The 2010 seminar included such topics as using social media tools for search engine optimization; and online reputation management. Social media tools discussed included Facebook, YouTube, Twitter and LinkedIn. It also included a panel of experts from retailers, interior designers and companies in other industries who had been successful in social media marketing.

The 2011 seminar included Brian Thornfelt, a key member of the management team at internet marketer Wayfair, Inc. Mr. Thornfelt cast the big picture of internet and mobile marketing. He talked about retailers and manufacturers who had been successful in using social media marketing and how to use blogging to engage customers.

The 2012 seminar covered subjects like the link between online content and commerce, inspiration and income. Other areas covered included return on investment (how to measure the effectiveness of online marketing in retail and interior design), the emerging power of online platforms like LivingSocial and Groupon and understanding the business model that uses QR codes.

Partnering with the "Next Generation"

One of the subjects I have heard periodically during the time I have been associated with the furniture industry is *next generation*. This term has been used by both manufacturers and retailers to consider the matter of succession, especially in family-owned-and-operated businesses. And along with thinking about succession comes the question, "What is the best way to prepare the sons and daughters of the company's owners and managers to be the next leaders of the company?"

It has been my privilege to know a significant number of families who have chosen to send members of their "next generation" to High Point College/ University as a way for them to get the training and industry exposure they will need as they prepare to take on meaningful roles in the family business. Although this is by no means a complete list, let's look at a few examples of members of the "next generation" who have graduated from High Point College/University and are making a difference in their family businesses.

BILTRITE Furniture—Leather—Mattresses: Fourth Generation

First, let's look at a unique example of High Point University partnering with the next generation of a retail furniture store. One phrase that I hear periodically is "stay local, buy local." This is something that seems to resonate with a significant number of retail furniture customers. BILTRITE Furniture—Leather—Mattresses, a retail furniture store in Milwaukee, Wisconsin, that has been family owned and operated since 1928 is a retail furniture store that effectively uses this philosophy. It is also a company that is proud of its history, and as a result, the owners decided to send two members of the fourth generation of the furniture family to High Point University to better prepare them to carry on the family tradition.

Randi (Komisar) Schachter graduated from High Point University in 2001 with a degree in interior design and a minor in art. Following graduation, she returned to Milwaukee to join the family business. Her younger brother Brad followed her to High Point University and graduated in 2004 with three majors: home furnishings marketing, home furnishings management and business administration. After graduation, he, like Randi, returned to Milwaukee to join the family business. Both Randi and Brad feel that their time at High Point University greatly helped them to be able to return to the store and be productive members of the BILTRITE team. Brad is very firm in his insistence that being in High Point, with the ability to experience the industry through activities such as field trips, was very beneficial.

Randi and Brad are now key members of the management team at BILTRITE. After being in the same location for fifty-seven years, in 2005, the company bought land in Greenfield, Wisconsin, where they built a new store. Randi, especially, played a huge role in designing the new showroom and making the move possible. A look at the BILTRITE organizational chart reveals that both Randi and Brad fill different but very complementary roles in ensuring the success of the organization. Brad is now the head of computer and warehouse operations, which involves everything from ensuring that orders are entered properly, to properly receiving merchandise from the store's vendors, storing goods in BILTRITE's state-of-the-art

(*Top to bottom*) Fourth generation of furniture retailer BILTRITE Furniture—Leather—Mattresses, Randi Komisar Schachter (known as Randi K.), a spokesperson for Biltrite; and fourth-generation, Brad Komisar, the head of computer and warehouse operations, and his wife, Sarah. *Courtesy of Biltrite Furniture—Leather—Mattresses.*

warehouse, retrieving sold merchandise, scheduling and delivering customer purchases and taking care of any customer service issues that may arise.

Randi, on the other hand, is the spokesperson for the store (referred to on the website, online and on-air as Randi K.) and is actively involved in the sales and marketing part of the business. She is the sales manager and does retail selling, floor displays, advertising and social media. It is obvious that both Randi and Brad wear many hats in ensuring that the customers have a good experience when doing business at BILTRITE. As an example of a family-owned-and-operated business, Brad's wife, Sarah, is also a full-time employee. She is a retail salesperson who trains new sales consultants and assists with store design projects.

Although both Randi and Brad have been successful, Randi's role in the success of BILTRITE led to her being recognized by *Home Furnishings Business* magazine as one of the "Forty Under 40" for 2018. This means that she was named as one of the forty most outstanding members of the home furnishings industry under the age of forty. The magazine describes these forty individuals as "passionate, successful individuals who do not let anything stand between them and their vision."

Kincaid Furniture: Fourth Generation

Now, let's look at an example of partnering with the next generation on the manufacturing side of the industry. Maxwell "Max" Kincaid Dyer is the fourth generation to be involved with Kincaid Furniture Company in Hudson, North Carolina. Founded through a combined effort of his grandfather and his great-grandfather, Kincaid Furniture Company is a great example of a very successful family-managed manufacturing business. Over the years, Kincaid Furniture has been known as a leader in solid wood bedroom, dining room and occasional furniture.

In 1988, La-Z-Boy Inc. purchased Kincaid Furniture as a way of expanding into the case goods market. Regardless of the fact that it is owned by a larger corporation, Kincaid Furniture has prospered and maintained its own brand identity in the marketplace.

Thinking he wanted to go into the family business, Max transferred from Appalachian State University to High Point University at the beginning of his junior year. In talking with Max, he said transferring to HPU was one of the best career-related decisions he ever made. His education at High Point University gave him the insight he needed to "hit the ground running" when he graduated in 1998 and returned to join Kincaid Furniture.

Max has always been on the sales and marketing side of the business. His love of the industry and Kincaid Furniture is obvious when watching him interact with retailers who visit Kincaid's High Point Furniture

High Point University graduate Max Dyer, fourth-generation Kincaid Furniture Company.
Courtesy of Richard Bennington.

Market showroom. Since Max joined Kincaid Furniture following his 1998 graduation from High Point University, he has been a key player in the management of the company. His success is reflected in the fact that the management of La-Z-Boy Inc. has promoted Max to his current position of vice president of sales for the case goods division. Today, La-Z-Boy case goods includes the Hammary and American Drew brands in addition to Kincaid Furniture.

Turner Furniture: Fourth Generation

High Point University graduate Russell Turner, fourth-generation Turner Fine Furniture. *Courtesy of Turner Fine Furniture.*

Scott Russell Turner Jr., or "Russell," came to High Point College for the opportunity to take courses in the furniture marketing program. Following his graduation from the college in 1989 with a double major in business and furniture marketing, Russell took a position with Lane Furniture. After two years with Lane, he returned to his family's business, Turner Fine Furniture in Thomasville, Georgia, where he was the fourth generation to have worked in the business. At that time, Turner Fine Furniture consisted of two relatively small retail furniture stores in the Thomasville, Georgia area.

In 2000, it was decided that Turner Fine Furniture would change its business model and become a holding company that owns and operates franchised Ashley Furniture HomeStores. Since that time, Russell has played a key role in expanding this organization into a network of sixteen Ashley Furniture HomeStores in six states: Georgia, North Carolina, Alabama, Mississippi, Florida and Virginia. These stores are supported by four company-owned distribution centers. This organization has experienced phenomenal growth, resulting in it being ranked seventy-fourth in *Furniture Today*'s 2018 ranking of the top-one-hundred retail furniture stores in the United States. Russell's personal contribution to the organization is recognized by his currently serving as CEO of the company, which is now listed as Russell Turner Furniture Holding Corporation.

Chapter 14

THE PROOF IS IN THE PUDDING

High Point University Graduates are Making a Difference in the Industry

N ow, let's look at some High Point University graduates who are not from furniture families. A saying I have often heard is "the proof is in the pudding," which is another way of saying "results matter." In this case, I would interpret this phrase to mean how well students who have attended High Point University are doing when they secured employment in the home furnishings industry after graduation says a lot about the university. Throughout this book, I have introduced you to former High Point University students who are successful in the industry, but I would like to close with a snapshot of two recent graduates and three others who have a relatively long tenure in the home furnishings industry and have been successful in their careers.

Let's start with two recent High Point graduates and take a brief look at their backgrounds and what they are doing in their careers.

MELISSA (DORSCH) EURILLO IS USING TECHNOLOGY TO DESIGN AND PRESENT PRODUCTS

Melissa (Dorsch) Eurillo came to visit High Point University from Red House, West Virginia, thinking she wanted to major in accounting and possibly become a CPA. When she came for a campus visit, she found out about the home furnishings marketing program and the High Point Furniture Market. Those two discoveries made her rethink everything. She started taking home

High Point University graduate Melissa (Dorsch) Eurillo, the director of marketing for Northridge Home. *Courtesy of Michelle Eurillo.*

furnishings–related courses and confirmed her choice of a home furnishings–related career through completing two internships during her college career at High Point University. She completed a marketing internship with Edward Ferrell/Lewis Mittman, a contract furniture manufacturer in High Point, and a sales and design internship with Furnitureland South. She graduated from HPU in 2011 with a degree in home furnishings marketing and a minor in graphic design.

Melissa's first job after graduation was working with Precedent Furniture, a division of Sherrill Furniture that produces custom-order upholstered furniture in Newton, North Carolina. After gaining furniture industry knowledge at Precedent, she moved on to work for an e-commerce company with retail clients such as Lowes and Home Depot. Eventually, she returned to High Point to work with Home Meridian International, an international sourcing company, where her position was director of digital imaging. In this position, she was in charge of 3-D modeling, primarily for their Accentrics Home Division, which offers a wide range of products for the e-commerce market. The products she helps design using 3-D modeling are then sourced from factories in Vietnam, China, Malaysia and India.

The 3-D modeling program Melissa used greatly reduced the need to build physical samples because she was able to use a wide variety of styles, fabrics and finishes to see many different options on a computer screen. This computer technology allowed her to see a 360-degree view of a product as well as observe how it would look in a variety of settings, such as being decorated for various holidays. These graphics can also be used by Home Meridian's dealers to show products on their web sites.

Recently, Melissa left Home Meridian to become director of marketing for a relatively new furniture company, Northridge Home. At Northridge Home, she oversees all photography, social media and the company's website. She is still involved with graphics and manages the graphics for key customers like Costco, Sam's and Lowe's Home Improvement.

Melissa feels it was a great advantage to combine the study of home furnishings with the study of graphic design. At High Point University, she found classes that had real-life applications to be very important. She recommends that current and future HPU students seek out internships and work at the High Point Furniture Market.

MEREDITH MATSAKIS IS USING HER VISUAL MERCHANDISING DESIGN MAJOR TO ACHIEVE SUCCESS IN NEW YORK CITY

Meredith Matsakis chose to attend High Point University because of its location near the High Point Furniture Market. During her four years at HPU, she took advantage of many different opportunities to learn about the home furnishings industry. She worked every furniture market and for premarket in showrooms like Calvin Klein Home, Bernhardt, Miles Talbott and Capris Furniture. She also had an internship in visual merchandising at Furnitureland South and two summer internships in New York City—the first with Design Works International and the second at the Kate Spade corporate offices.

Meredith graduated in 2017 with a degree in visual merchandising and a minor in theater performance. During her senior year, she conducted an independent study with the High Point University Communications Department. For her independent study, she wrote and produced five video episodes of "Millennials and the Furniture Industry," which underscored the importance of lifestyle marketing, especially in regard to conspicuous products like home furnishings.

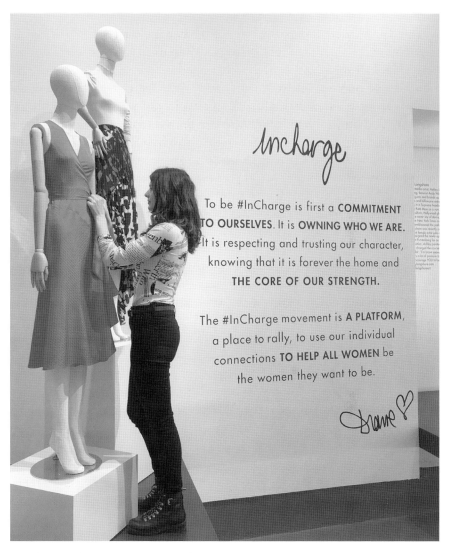

High Point University graduate Meredith Matsakis, working as a visual merchandising manager for Diane von Furstenberg. *Courtesy of Meredith Matsakis.*

After graduation, she took a job as studio content manager with Design Works International in New York, the location of her first New York City summer internship. Design Works International is a lifestyle studio that works with various clients on product development, trends forecasting, graphic design and digital textile printing. In her job as studio content manager, she was responsible for creating licensed brands and telling lifestyle stories for companies in the home furnishings industry that are marketing such products as rugs, pillows, wall art, bedding and fabrics. Meredith felt that her major and minor helped her get her job with Design Works International and that her independent study with the High Point University Communications Department gave her the confidence she needed to be successful in her job.

After gaining valuable work experience with Design Works International, she has taken a position with design icon Diane von Furstenberg (DVF) as corporate visual merchandiser in the DVF corporate office in New York City. Meredith feels that she is now using what she learned in her visual merchandising classes at High Point University. She works directly with the company's director of visual merchandising on the implementation and execution of window concepts, in-store presentation, showroom set up, brand events and pop-ups. She is responsible for creating visual merchandising guidelines and ensuring that all DVF stores and partners have the necessary tools and guidance to maintain company visual merchandising standards. Additionally, she travels to the stores to train the brand managers. She says that her goal as corporate visual merchandiser is to continuously elevate the consumer experience through visual design.

MORE VETERAN GRADUATES IN THE INDUSTRY

Finally, let's take a brief look at three graduates who have been in the industry for a number of years. A look at their career paths will provide insight into how they have been successful, even though each has a different industry-related career.

Marc Sholar, a 1991 High Point University graduate, enlisted in the merchant marine after high school. After four years in the merchant marine, he briefly attended a community college near his home but transferred to High Point University, where he graduated with a degree in home furnishings marketing. Having developed an interest in sales, he took a job as a sales representative for Stanley Furniture Company. After eight years, he left Stanley to become a sales representative with Universal Furniture Company, where he is

High Point University graduate Marc Sholar, a sales representative for Universal Furniture. *Courtesy of Richard Bennington.*

High Point University graduate Erica Wingo, the vice-president of merchandising and marketing at Baker Interiors. *Courtesy of Richard Bennington.*

employed today. Marc feels that the product knowledge he gained in his home furnishings classes were very important to him becoming a successful sales representative. He believes the class presentations by professionals from the industry were especially valuable to his understanding of the industry and how companies position their products in the marketplace. He is happy with his career and credits a considerable part of his success as a manufacturer's sales representative to his studying home furnishings marketing at High Point University.

Erica Wingo, a 2007 graduate, came to High Point University because of its home furnishings–related majors. She had become interested in a career in the home furnishings industry because her father was the manager of a rather large retail furniture store in her hometown of Farmville, Virginia. At High Point University, she began the interview process during her junior year, which she thinks gave her the confidence she needed to get a job and be successful. She also completed an internship with La-Z-Boy Inc. while she was at High Point University.

After graduation, with a degree in home furnishings marketing and a minor in business administration, she took a job in product development with Hooker Furniture Corporation. During her time with Hooker, she moved from management trainee to manager of merchandising and product development and eventually became merchandising director for the company. She was successful in product development, which is evidenced by one of her Hooker Furniture designs being so innovative that it received a Pinnacle Award from the American Society of Furniture Designers.

She left Hooker Furniture to become senior director of merchandising at Bernhardt Furniture. Again, one of her Bernhardt designs received a Pinnacle Award. Eventually, Erica left Bernhardt to assume her current position of vice-president of merchandising and marketing at high-end manufacturer Baker Interiors Group. She is in charge of product development, which includes sourcing from all over the world. This position also includes fabric selection and manufacturer exclusive finishes for the Baker and McGuire brands. She is also responsible for merchandising their Milling Road–branded merchandise.

The last graduate I would like to introduce is David Koehler. David came from Delaware to High Point University with the idea of majoring in business administration and economics. Once he arrived in High Point, he was able to get a part-time job setting up showrooms for the furniture market. His job involved opening boxes and helping arrange the merchandise in the showrooms. This job got him interested in knowing more about the furniture industry, so he took furniture-related courses as electives in his major. His part-time market job and the High Point University home furnishings courses made him interested in a furniture-related career.

When he graduated in 1984, he was hired and went into the training program of Atlanta-based retail furniture chain Havertys Furniture

Company, where he was employed for twenty-one years. During the time he was employed with Havertys, he was transferred several times. Some of the positions he held were assistant vice-president of operations and regional vice-president in Texas, where he was in charge of all the Havertys Furniture stores. He also opened the northern Virginia market for the company.

Eventually, David left Havertys to become regional vice-president of sales for La-Z-Boy Inc. Shortly after joining La-Z-Boy, he had the opportunity to become the CEO of top-one-hundred furniture retailer Johnny Janosik with headquarters in Laurel, Delaware. David decided to take this position and is chairman and part-owner of the company today.

A look at these graduates and others mentioned throughout the book is proof that the home furnishings–related classes, help them be more successful in their industry jobs whether these jobs are in sales, marketing, management, design or visual merchandising.

The challenge today is how to keep up in the ever-changing world of home furnishings. This begins with studying today's consumers, their personalities; where they live, which includes region of the country, whether they live in rural, suburban or urban settings; what they expect in the products they buy, both functionally and aesthetically; where they buy in terms of online versus physical store locations; how they prefer to pay for the products; and many other factors.

High Point University graduate David Koehler, chairman and part-owner of Johnny Janosik Inc. *Courtesy of David Koehler.*

To adequately prepare students to go out into the home furnishings industry, the university must continue to develop and maintain a working partnership with professionals in the industry. High Point University is in a unique position because it is located in the furniture capital of the world. This partnership will be even more important in the future. It is essential to continue to bring the real, ever-changing industry into the classroom through guest speakers, field trips to area home furnishings companies and visits to the furniture market. Students must also continue to learn from internships, and working at the market or with local stores, manufacturers, distributers and others.

But what are the prospects for future partnerships between High Point University and the furniture industry? Read chapter 15 and find out.

Chapter 15

CONCLUSION

The Future Is Bright for a Mature, Vibrant, Ever-Changing Partnership Between High Point University and the Furniture Industry

At the beginning of this book, I mentioned that the development of a partnership between the furniture industry and High Point University—the synergy between "town" and "gown"—is an unusual journey deeply rooted in a shared history spanning nearly a century. It is a journey of mutual growth built on a foundation of common values. The journey is not over yet, and in some ways, the future may be more exciting than ever.

What will the partnership between the furniture industry and High Point University be like in the future? No one can say for sure, but one thing is certain: the future is bright for a continuing, vibrant, ever-changing partnership between High Point University and the furniture industry.

High Point University is indeed fortunate to have such a forward-thinking president as Dr. Nido Qubein, who frequently refers to High Point University as High Point's University. He has been quoted as advocating partnerships through which the university and the city of High Point will thrive. In fact, he has recently put together a town and gown think tank to evaluate ideas that will mutually benefit the city of High Point and High Point University. In recognition of the importance of the furniture industry, and especially the furniture market, to the city of High Point, the High Point Market Authority president and CEO was named to that board. Since High Point is widely known as "the furniture capital of the world," there are many ways future partnerships may develop. For example, a logical area is with the world's largest furniture market the first thing that is usually thought of by people when they hear the name High Point.

Another direction the partnership might take is with graduates joining manufacturers to help them develop innovative products or ways of selling products. What could be more promising than a positive partnership between High Point University, which has been ranked by *U.S. News & World Report* as the number one most innovative university in the South for the past four years, and the furniture industry, where success thrives because of innovation?

Yes, a chair is a chair, a table is a table, a sofa is a sofa and a bed is a bed, but there is far more to it than that. Home furnishings are functional art forms, which means that they have a particular style in addition to providing a basic function in the home. At a seminar at the High Point Market, I once heard the late Martin Roberts, a retail store designer from New England, say that home furnishings are "tools for living." This means that function is important but so is the lifestyle of the purchaser, the makeup of the family and many other variables that affect the type of furnishings that are best for a particular family situation. Although lifestyles and family buying preferences have been researched extensively in the past, marketers today cannot depend on past research alone. Successful marketers must keep up with the latest consumer needs and, at times, do their own research to understand and reach the ever-evolving consumer market.

Throughout this book, I have used the phrase, "The times, they are 'a-changing.'" This is definitely true throughout the furniture industry as well as on college campuses. Although furniture is a product everyone needs, where people buy, how they buy and what they buy are all changing. And people are not all the same. The overall market for furniture is in a constant state of change. But successful marketers must find a niche, or target market, where the prospective customers have a need for the product, sufficient purchasing power and the market is of sufficient size for the company to generate a profit. They must then develop a marketing strategy to reach their chosen target market, which takes into consideration that not all customers shop alike. Some shop online, others shop in furniture stores and still others choose to buy from general merchandise stores where home furnishings is just one category of products they offer. And where they look for information about furniture also varies.

When I first started learning about the furniture industry in the late 1970s and early 1980s, the preferred method of shopping for home furnishings products seemed to be through looking at advertisements in the local newspaper. But today, it seems that fewer and fewer people read newspapers, let alone look at the advertisements. Television, with networks like HGTV, is having an impact. Websites and social media platforms like Facebook, Instagram and Pinterest are having an increasing impact, especially with millennials and generation Z consumers.

Lifestyles are changing; technology is affecting the industry; the way consumers shop is changing; and this is just the "tip of the iceberg." Where does High Point University fit in? The fact that the university is located in the "furniture capital of the world" gives us a hint. As I pointed out earlier, this proximity provides a natural tie-in for a continuing partnership between "town" and "gown."

The Industry Wants a Continuing Partnership

There is evidence that the industry wants to continue to partner with High Point University. For example, Bassett Furniture Industries has used an actual, real-world project as a way of obtaining ideas from High Point University visual merchandising design and interior design students for selected areas of their retail stores. The project involved having junior-level High Point University students work in groups to design a retail showroom to effectively present selected Bassett products to young families with preteen and teenage boys and girls. The students were given specific constraints,

High Point University students presenting their ideas for a juvenile furniture section in a Bassett Furniture Store. *Courtesy of Cathy Nowicki.*

including square footage of the showroom, color palettes and furniture collections from Bassett.

The students presented their ideas to a panel of representatives from Bassett who then gave them instant feedback. This was a win-win situation for both Bassett and the students. Bassett was able to get possible showroom designs as seen through the eyes of young people, and the students got a professional critique, which allowed them to see how their designs would likely be perceived by consumers shopping in Bassett stores. And often, the students appreciate the opportunity to have professionals critique their designs. In the words of interior design major Aria Real, "It was such a grand opportunity to present to the Bassett representatives. I gained confidence in my presentation skills and in my abilities as a designer."

STEELCASE EXPERIMENTAL INTERACTIVE CLASSROOM

Another example of the industry partnering with High Point University was in 2019, when Steelcase, a large manufacturer of contract furniture, selected High Point University to receive a two-year grant to study the effect that furniture and space has on student learning. Under this

Steelcase experimental active learning classroom in Norton Hall. *Courtesy of Richard Bennington.*

research partnership, Steelcase has provided a variety of exciting, colorful furnishings, including individual mobile and configurable desks where students can work individually or in teams. There are also other desks where the students can stand and work. Individual whiteboards are provided for the students to use when they are doing classwork or perhaps team projects. These furnishings are designed so students can take ownership of the classroom. This is designed to give Steelcase helpful feedback about product features that should be in the finished products they decide to introduce to the potential customers in the marketplace.

Partnering with the Market

Another example of a High Point University student partnering with the industry was when Allison "Addie" Gantt, who you were introduced to in chapter 5, was chosen by International Market Centers during the fall semester of her senior year to participate in their "Fresh to Follow Instagram Takeover." For this takeover, which was for the fall 2018 High Point Furniture Market, Addie was assigned a section of the market to look for what she thought were fresh new ideas. In her assigned section, which was the Suites at Market Square, she was asked to record what caught her eye, where it was, who it was made by and what she liked about it. Cindy Hodnett, director of public relations and communications for International Market Centers, then interviewed her and put the interview on Instagram. This is a good example of the furniture industry attempting to determine what will catch the eyes of generation Z.

I HOPE YOU HAVE enjoyed our journey through the history of the partnership between High Point University and the furniture industry. During this journey, I wanted you to meet some of the many students that have taken the home furnishings–related classes. It is interesting that some came to High Point because they were interested in home furnishings–related careers but that others discovered the career possibilities after they arrived in High Point. They may have worked during one or more furniture markets, or perhaps taken some home furnishings–related classes as electives. Regardless, they found a career that has become very satisfying for them.

TO SUMMARIZE, OPPORTUNITIES ABOUND

A lot has been written about how the furniture industry has moved from being primarily manufacturing-oriented to being marketing-oriented. But that doesn't mean that there are not still opportunities for activities related to furniture manufacturing available in the High Point area. For example, classes have learned about wood furniture manufacturing through field trips to Marsh Furniture to see custom wood kitchen cabinets and other built-in components being manufactured in their High Point plant. Another wood furniture factory that students have visited is Davis Furniture, where they were able to see the manufacturing of wood office furniture. Students have also seen the actual manufacturing of upholstered furniture through field trips to the Miles Talbott and Thayer Coggin furniture factories. These are only two of several upholstered furniture factories located in the High Point area.

Although the primary focus in High Point is usually on the High Point Furniture Market and local furniture plants, there are other businesses in the area where High Point University students can learn about marketing home furnishings. For example, students have visited local photography and marketing companies like Kreber Inc. These are the companies that create the visual images that are so important in the marketing of home furnishings.

As discussed in chapter 13, another company often visited by students is Furnitureland South, the world's largest furniture store. The partnership with Furnitureland South exposes students to the retail side of the industry. It is unique for students to have access to such a wide variety of retail store displays showcasing the store's five hundred lines of home furnishings products and a world-class, state-of-the-art furniture distribution center, all in one location just a short drive from the High Point University campus.

Of course, the focus of many High Point University students is the High Point Furniture Market. The High Point Furniture Market has moved from its origins of being a regional showcase for predominantly lower-priced furniture to being the largest, most comprehensive wholesale market for home furnishings in the world, with a total annual economic impact, according to a recent study, of $3.59 billion. It is a place where buyers from many different types of retail home furnishings stores and other resellers can shop for the specific furnishings they think will sell in their stores. In addition, the High Point Market is promoting itself as the place where a broad cross section of the creative minds behind the products meet and talk with market visitors. This is a great advantage for retailers and designers when they interact with consumers who want to know the story behind the products they are thinking of purchasing. And of course, this is an outstanding educational resource for High Point University students.

Since its beginnings almost one hundred years ago in the city of High Point, High Point University has moved from being a relatively small, church-related college to being a world-class, innovative university that prepares students for careers in the real world—as High Point University president Nido Qubein has said, not as these careers have been in the past but how they will be in the future. Both the furniture industry and High Point University now find themselves in a world economy. Looking from the viewpoint of High Point University, how better to prepare students to be the leaders for tomorrow than through exposure to the High Point Market, where they interact with people and products from all over the world? And it is very unusual for students to have an opportunity to learn how the High Point Market Authority, International Market Centers and individual market exhibitors have come together to create a world-class experience for retailers, designers and other market visitors.

Students with widely varying majors are already involved at the market in various capacities. Majors from the High Point University School of Art and Design are exposed to a real-life example of a design-oriented industry in action. Whether the students are interested in product design, interior design, graphic design, visual merchandising or fashion merchandising, it is a great learning experience.

In addition to being a design-oriented industry, the home furnishings industry is an entrepreneurial industry. Therefore, what better place is there for students majoring in entrepreneurship to gain real-world knowledge than from the entrepreneurs in the home furnishings industry? Majors from the business school are gaining real-world knowledge in many ways. Students who are interested in other areas, such as supply chain management, marketing, and sales, also have a unique opportunity to see actual examples of people and companies participating in the High Point Market.

A wide range of other majors can also benefit. Event management majors can see firsthand how the furniture market, itself a very large trade show, is organized and managed successfully. And they can see what is involved in managing many of the parts and pieces that are required to fit together for the market to run smoothly. Such parts and pieces include things like registration, transportation, a press center and specific events like educational seminars and banquets must be carefully managed for the market visitors to get the most out of their trips to market. Communications majors can get a feel for how the knowledge and skills acquired in their major can be used in real-world situations. Foreign language majors can help with foreign language interpreting because of the large international presence at market. And the list doesn't stop there. Because of the size and complexity of the market, students with many majors can benefit from the experience.

From the viewpoint of the market and the industry, it is very advantageous to continue to partner with High Point University and, in many cases, expand the existing partnership. Many advantages, like the need for people to staff the market and access to young people with creative minds have already been mentioned, but the need of the industry for top-notch, well-trained, forward-thinking graduates to assume the jobs of tomorrow is another reason for the partnership. Trade publications are publishing articles focusing on how successful companies are using modern management and marketing techniques, like analytics, to base their decisions on actual data analysis to gain a competitive edge. This will translate to an even greater need to hire bright young people with the knowledge of these twenty-first-century management tools. And when graduates have this knowledge combined with an understanding of how the home furnishings industry works, they will be in even higher demand.

I predict that the partnership will remain strong in the future. It is a win-win situation for the furniture industry and High Point University to maintain strong bonds. High Point University is High Point's university, and High Point, North Carolina, is the most recognizable furniture city anywhere. They are strong partners, and their relationship is tangible evidence that it is possible for a progressive university and a receptive local industry to form an effective and mutually beneficial partnership.

INDEX

ABOUT THE AUTHOR

 native of Virginia, Richard R. Bennington joined the faculty of High Point College, now High Point University, in August 1974. He earned a bachelor's degree in economics and business from Emory and Henry College and a master's degree in business administration from Virginia Tech. After three years of teaching at Averett College, now Averett University, in Danville, Virginia, he enrolled at the University of Georgia, where he earned a doctorate in business education. At the University of Georgia, he was awarded the Outstanding Graduate Research Award by Delta Pi Epsilon Professional Fraternity in Business Education. In 1974, he came to High Point College/University, where, over time, he has served as a professor of business, chair of the Earl N. Phillips School of Business, director of the Home Furnishings Marketing Program and a Paul Broyhill professor of home furnishings. He has been awarded the Meredith Clark Slane Outstanding Teaching/Service Award (historically High Point University's most prestigious faculty award). He is the author of the book *Furniture Marketing, from Product Development to Distribution*, published by Fairchild Books, a division of Bloomsbury Publishing in New York City. He has traveled extensively throughout the furniture industry in the United States and has spoken and done projects within the industry both in the United States and abroad.